Plant Biosecurity Policy Evaluation

The Economic Impacts of Pests and Diseases

Plant Biosecurity Policy Evaluation

The Economic Impacts of Pests and Diseases

David Cook
University of Western Australia, Australia

Rob Fraser
University of Kent, UK

Andrew Wilby
University of Lancaster, UK

World Scientific

NEW JERSEY • LONDON • SINGAPORE • BEIJING • SHANGHAI • HONG KONG • TAIPEI • CHENNAI • TOKYO

Published by

World Scientific Publishing Europe Ltd.
57 Shelton Street, Covent Garden, London WC2H 9HE
Head office: 5 Toh Tuck Link, Singapore 596224
USA office: 27 Warren Street, Suite 401-402, Hackensack, NJ 07601

Library of Congress Cataloging-in-Publication Data
Names: Fraser, Robert W. | Cook, David (David C.) (Agricultural economist) | Wilby, Andrew.
Title: Plant biosecurity policy evaluation : the economic impacts of pests and diseases /
    by Robert Fraser (University of Kent, UK), David Cook (University of Western Australia,
    Australia), Andrew Wilby (University of Lancaster, UK).
Description: New Jersey : World Scientific, 2017. | Includes bibliographical references.
Identifiers: LCCN 2016038677 | ISBN 9781786342157 (hc : alk. paper)
Subjects: LCSH: Agricultural pests--Economic aspects. | Plant diseases--Economic aspects. |
    Introduced organisms--Control. | Nonindigenous pests--Control. | Plants, Protection of.
Classification: LCC SB601 .F73 2017 | DDC 632/.6--dc23
LC record available at https://lccn.loc.gov/2016038677

British Library Cataloguing-in-Publication Data
A catalogue record for this book is available from the British Library.

Desk Editors: Anthony Alexander/Mary Simpson

Typeset by Stallion Press
Email: enquiries@stallionpress.com

Printed in Singapore

*To the extent that risk analysis is precise
and simple, it is not real. To the extent that risk analysis
is real and complex, it is not precise*

(Haimes, 2009).

FOREWORD

Much of my day job for the last several years has involved applying the ideas and principles laid out in this book by Dave Cook, Rob Fraser and Andy Wilby, to plant health issues in the UK and Europe. The threat to plant health in the UK is a microcosm of the global threat of invasive species. And, it is a threat that continues to grow, as evidenced by the upward trend in invasive species like tree pests and pathogens entering since 1900.

The authors' understanding goes beyond the formalisation of a framework for assessing biosecurity preparedness and incursion response policy in that they clearly appreciate the pressurised situations within which decisions are made. Their description of the decision process certainly struck a chord with me and will, I suspect, have the same effect on many colleagues, acquaintances and friends at the sharp end of decision making in this area:

> *Virtually all decision support people are time-pressured. To influence private or government action our methods of calculating benefits and costs must be expedient. Biosecurity staff seldom have the luxury of researching specific species in detail over months or years. Instead, they are usually asked to predict the economic, environmental and social impacts of threatening or newly arrived species in areas they have not been observed in before; all within a matter of hours, days or (at best) weeks. Time is critical, particularly in the case of new species incursions, because the window of opportunity for successfully removing them is usually short. Moreover, the context to which a response effort is to be made constantly changes due to external pressures, like political and economic cycles.* (p. 5)

The clear and strong understanding of the policy context, and the conditions under which decisions to prevent, eradicate, control, and adapt are made, flows through the chapters that are laid out. This makes their work accessible to a wide range of potential readers and users. It will be of interest to all the different disciplines involved in managing plant biosecurity as well as will lend itself well to teachers and students in agriculture, economics, business studies, and environmental management, amongst others.

The book recognises and explains some of the key complexities of plant biosecurity and offers accessible and clear ways to tackle them. It is not uncommon for decision makers to be faced with a biosecurity threat with limited information and significant areas of uncertainty — gaps that are unlikely to be filled before a response is required. The flexibility of this tool kit for decision makers is therefore highly appropriate.

The methods put forward allow complexities to be considered in a transparent fashion; the importance of which cannot be understated when explaining or justifying a particular biosecurity response. Of particular interest in the UK is the impact of pests and diseases of trees and the associated non-market values affected. Such impacts illustrate the importance of the time horizon of the analysis (with sometimes long lags before the full losses become apparent) and the need to incorporate more than relatively simple market costs. The participatory technique for including a broader set of criteria into decision making, while subject to a number of shortcomings, does allow for transparent communication of the impacts of the pests and diseases as well as examining, testing and sharing probabilistic outcomes of particular decisions.

The deliberative multi-criteria decision-facilitation approach allows the incorporation of diverse and context-specific knowledge from a wide range of stakeholders. This seems a particularly apposite option in the face of increasing biosecurity threats at a time of pressured plant health budgets. Given these pressures, there is a pressing need for such deliberative approaches to be harnessed to proactively prepare the ground for policy change and gain more public trust and credibility.

The enhanced communication of biosecurity responses that this book describes could aid in the sharing of responsibilities across public and private domains, and in justifying response costs. It provides a rationale for making improved plant health policy decisions for both present and future societies, but acknowledges that those decisions might never be perfect.

Glyn Jones

Environmental Economist, Fera Science Ltd.
National Agri-Food Innovation Campus
Sand Hutton, York YO41 1LZ, UK

glyn.d.jones@fera.co.uk

ABOUT THE AUTHORS

David Cook is Adjunct Associate Professor of Agricultural Economics at the University of Western Australia School of Agriculture and Environment (UWA-AE). He completed an honours degree in Economics at Murdoch University in 1995 and a PhD (Agriculture) at UWA-AE in 2001. He has worked as postdoctoral researcher at Imperial College London (2003–2004) and as a Research Economist with CSIRO Sustainable Ecosystems (2004–2011) and the Department of Agriculture and Food, Western Australia (1996–2004, 2011–). He hails from Katanning in the Great Southern region of Western Australia, an area where broad acre agriculture is the dominant land use and driver of the region's economy. As a child, he developed a fascination with the natural world he has never lost, and has taken this enthusiasm into a career studying what happens when species of plants and animals are introduced to new agricultural and ecological systems. Since 2002, he has authored over 40 peer-reviewed papers in a diverse range of ecology, economics and policy journals and well over 100 cost–benefit analyses and impact assessments for different tiers of government. This work has concerned pests and diseases from all taxonomic groups, and involved both conventional and unconventional impact assessments and decision-facilitation techniques.

Rob Fraser is Emeritus Professor of Agricultural Economics. He completed a first degree in Economics at Adelaide University before gaining a Rhodes Scholarship to study for MPhil and DPhil qualifications in Economics at Oxford University from 1978. His first appointment was as an Assistant Professor of Economics for the University of Virginia, USA in 1981. A year later, he joined the University of Western Australia initially as a Lecturer in Economics, becoming Professor of Agricultural and Resource Economics early in 1999. In 2000, he was appointed Professor of Agricultural Economics at Imperial College, and in 2006 he joined the University of Kent in the same capacity.

He has an international research reputation as a policy economist, specialising in both agri-environmental and invasive species policy design and evaluation. In this context, since moving to the UK in 2000 he has participated in a range of DEFRA and other funded research projects. In addition, he was commissioned in 2006 by the OECD to prepare a report on "Information Deficiencies in Agri-Environmental Policies", which was then presented as the keynote paper to an OECD Workshop on this topic. He is a Past President of the Agricultural Economics Society (AES) and is both a Past President and a Distinguished Fellow of the Australian Agricultural and Resource Economics Society (AARES). He is also a Member of the Editorial Board of the *Journal of Agricultural Economics.* Since 2012, he has been a Member of DEFRA's Economic Advisory Panel.

Andrew Wilby is Senior Lecturer in Conservation Biology at Lancaster University. He completed his first degree in Applied Biology at the University of Bath before studying towards a PhD in Ecology at Imperial College London, which was completed in 1996. He then completed a series of postdoctoral research projects at Ben-Gurion University of the Negev, CAB International and Imperial College London, before attaining his first appointment as Lecturer in Agriculture and Environment at the University of Reading in 2004. In 2007, he moved to a Lectureship at Lancaster Environment Centre, Lancaster University.

He is a community ecologist specialising in agricultural ecology. His research aims to apply ecological understanding to agricultural problems, such as the maintenance of natural pest regulation and pollination, and the sustainable integration of agriculture and biodiversity conservation. To this end, he has participated in a range of projects funded by of DEFRA, NERC, BBSRC as well as industry bodies, such as the AHDB, and is an Editorial Board Member of the *Bulletin of Entomological Research.*

ACKNOWLEDGEMENTS

The authors especially thank Professor Jeff K. Waage (London International Development Centre) and Professor John D. Mumford (Imperial College London) for engaging us in the Department for Environment, Food and Rural Affairs project *A New Agenda for Biosecurity* which provided the impetus for much of our subsequent research. We would like to thank the Department of Agriculture and Food, Western Australia; the University of Western Australia; the Commonwealth Science and Industrial Research Organisation; the Plant Biosecurity Cooperative Research Centre; Imperial College London; the University of Kent; the Department for Environment, Food and Rural Affairs; North Carolina State University; the United States Department of Agriculture; and the International Pest Risk Research Group. Their support for our ideas over the years has been tremendous.

CONTENTS

CHAPTER 1

BACKGROUND AND OBJECTIVES

1.1 Analysing Invasive Alien Species (IAS) Risk

This book acknowledges that we can never know enough about agro-environmental systems to maintain them perfectly; nor can we know enough to respond perfectly to changes in these systems induced by invasive alien species (IAS). IAS are organisms that have or could potentially have a net negative effect on societies and are capable of reaching new areas via natural or human-aided means. The debate about how best to control their impacts on society is a battleground of ideologies that on the one hand wants governments to pay for IAS management, and on the other wants private 'beneficiaries' to pay for it. We refer to activities related to the prevention and management of IAS damage as 'biosecurity'.

Biosecurity is plagued (please pardon the pun) by measurement problems. We know IAS can cause a lot of damage, but no amount of measurement or experimentation can tell us how large a problem it presents day-to-day, or year-to-year. Social, ecological and economic systems are, after all, complex.

Uncertainty within agro-environmental systems is a problem for both markets and governments alike. Instances have and will always occur where third parties are affected by economic activity, and there can be no better example of this than IAS damage. It represents a negative externality caused by industries and governments in the

---

Box 1. 'Invasive Alien Species' Defined

An IAS is a species that does not naturally occur in a specific region and whose introduction does or is likely to cause a net loss in social, economic or environmental welfare. Various terms have been used to describe or categorise problematic species, including *alien, exotic, invasive, noxious, non-indigenous, non-native, nuisance, pest* and *weed*. This array of adjectives has sometimes led to confusion and misunderstanding between scientists and policymakers. Use of the word invasive, in particular, has been problematic as ecologists often use it to describe species that spread quickly from an initial site of introduction, whereas in policy and legal documents it tends to imply negative effects caused to societies. A major source of confusion lies in the fact that a species' invasiveness does not necessarily imply it will cause large impacts (Ricciardi and Cohen, 2007).

---

business of moving host material from place to place, and agriculture, environment and society are the affected third parties.

Because it is difficult to measure, we cannot expect the price mechanism to include a social or environmental externality when balancing the supply and demand of transported goods. If the externality is negative, not including it necessarily means the price of the product concerned is too low, and demand for it is too high.

It is sometimes easy to overlook the fact that price does capture a plethora of other sophisticated information. In fact, price is so effective in conveying so much to markets that it is rather ominous that it cannot internalise IAS externalities all by itself. It hints at an imposing information problem; one that is also faced by government should it intervene in IAS-related issues.

We make no distinction between a decision support person working for a government agency or a private industry. Our interest is purely in IAS preparedness and response decisions. These are made on a regular basis by different tiers of government and private industry, and to the extent possible are informed by analytical information. In terms of economics, the most common form this information takes is cost–benefit analyses that determine net financial gains or losses from taking certain courses of action, be they responses

Box 2. What is a Negative Externality?

Son   *Hey Dad, have a look at the skid I did on the gravel by the path!*

Father   *Very impressive son, now go and clean it up.*

Son   *Why?*

Father   *Because there are now stones all over the path, which I often walk on to get the paper in the morning. As a result there is now an increased likelihood I will step on a stone and hurt my foot. In other words, while your play activity has given you a benefit, you have imposed a cost on me called a 'negative externality'. I may not experience that cost straight away; I might be lucky for a time and not tread on one of the stones. However, in the long run I have a much higher chance of incurring injury by treading on one. Now, if you had saved all of your pocket money and bought this house, you would have what is known as a property right. This would entitle you to do whatever you wish with the house, including doing skiddies on the gravel by the path. I would still be inconvenienced by this behaviour, but I could internalise the externality by offering you an incentive like increased pocket money to clean up the stones and not do any more skids. Unfortunately for you, I happen to own the property right and I require the path to be restored to its former glory, please; that is, unless you'd care to offer me compensation for the externality.*

Son   *Clayton had a nose bleed at school today.*

to incursions when they happen or risk-reduction measures before they happen.

As we shall see though, there are limits to what conventional cost–benefit analyses can be expected to deliver. Virtually all decision support people are time-pressured. To influence private or government action our methods of calculating benefits and costs must be expedient. Biosecurity staff seldom have the luxury of researching specific species in detail over months or years. Instead, they are usually asked to predict the economic, environmental and social impacts of threatening or newly arrived species in areas they have not been observed in before; all within a matter of hours, days or

---

Box 3. What is the Appropriate Role for Government in Biosecurity?

It is difficult to say categorically what biosecurity activities government should be involved in and what it should leave to the market. The authors neither advocate government intervention in all biosecurity matters nor discourage it. Instead we focus on the underlying principle that the contributions of public and private institutions to an intervention should match their relative share of benefits. The problem is that both governments and markets have inadequate information to identify beneficiaries or to internalise IAS externalities. Let's take the example of an imported good that could act as a pathway for a harmful IAS that can affect ecosystems as well as agricultural crops. While the price of the good includes a host of information about its producer's comparative advantage, it does not include intangible information about the potential damage caused if importing the good leads to an IAS introduction. There are no easily-accessible prices for non-market assets like native ecosystems, so the market cannot correct the negative externality associated with importing the good. By the same token, governmental intervention to correct the situation faces the same problem. Suppose a government chooses to impose a tax on the good in the form of costly IAS risk-reduction measures that must be undertaken prior to import. Without information on what the IAS might cost if it was introduced, the government cannot know the appropriate tax to impose. Even if it did, the tax would be inequitable as it would inflate the price of the good for all consumers, including those who receive a negligible share of any benefit created. The World Trade Organization (WTO) (itself a form of government influencing global markets) has rules and guidelines that Member countries must abide by when taxing imports, further complicating matters.

(at best) weeks. Time is critical, particularly in the case of new species incursions, because the window of opportunity for successfully removing them is usually short. Moreover, the context in which a response effort is to be made constantly changes due to external pressures, like political and economic cycles.

This book describes methods the authors have used to estimate the value of IAS externalities on agriculture and, when necessary, to combine them with non-monetary information concerning environmental and social externalities. The ideas put forward move away from scenario-based decision support analyses, concentrating instead on combining *probabilistic impact models* that show what might happen if species 'invade' a region and *values models* that help to determine what actions are taken given the possible impacts. This approach has evolved from the ground up, as it were, by informing biosecurity decisions predominantly in the State of Western Australia, but elsewhere around the world. This has been done with the help of both private and government institutions.

Examples are used to show how a decision support person might use probabilistic impact models to rapidly predict the economic effects IAS could have without taking too many liberties with regard to biology or economics. All scientific and economic concepts used in these models are relatively simple and well-documented, and when combined form a sufficiently detailed and defendable foundation for financial decisions. However, examples will also be shown where these models can be strengthened through the use of a values model that considers non-monetary components of impact that are also relevant to decision makers. This is an important requirement of any decision support method involving IAS of agricultural and non-agricultural significance.

The ideas put forward in the book are far from a panacea. They are practical and accessible, but are only a snapshot in time. Decision support systems are evolving, and in the course of our discussions we will mention several new developments set to revolutionise the way complex information is delivered to decision makers. Some of these are tantalisingly close to being used in practice. Indeed, the rapid development of interactive map-based technologies and agent-based "games" could make our snapshot age prematurely. We certainly hope so! However, we also hope these future tools can draw on our methods, build on them and remain adaptable to the complexities of agro-environmental systems as well as the constraints faced by decision support people the world over.

1.2 Agricultural vs. Environmental Systems: Regulatory Policy on IAS Externalities

Since you had cause to pick up this book in the first place, you are most probably aware that the threats posed by IAS are very real, potentially very damaging and are — now here is where we hit trouble — probably increasing, most likely as a result of trade. As globalisation continues, non-tariff trade barriers safeguarding movements of IAS-host material are being placed under increased scrutiny, and are sometimes difficult to justify given their inflationary impact on consumer prices. While the costs of non-tariff barriers are immediate and relatively certain, the benefits that accrue over time are highly uncertain and are subject to discounting which lowers their value as we move forward in time. Worse, the benefits may not be restricted to agricultural commodities with known values, and may also include environmental, social or cultural assets with unknown values.

Governments have long recognised the threat of global species exchange and it has been relatively easy to develop broad political consensus on IAS problems, as they constitute a shared, external threat which affects many economies. However, when it comes to the technical side of intervening to fix the IAS externality, governments have failed for the same reason that the market created the problem in the first place. There is insufficient information about potential IAS impacts to internalise the externality.

This lack of critical information means that government intervention to fix the problem has largely been underpinned by anecdotal, rather than analytical information about IAS impact and risk. With a large number of biological invasion "horror stories" to draw from, literature on IAS problems has raised awareness of the global nature of this problem (e.g. Bright, 1998; Baskin, 2002). Despite them being, in many cases, vague and non-specific, 'broad brush' quantitative economic assessments have been used by policymakers simply because there was no alternative. Perhaps most significant has been work by OTA (1993) and Pimentel *et al.* (2000), the latter famously estimating that the annual cost to the United States of IAS and their control was $137 billion. The US Executive Order of 2000, by which President Clinton established an inter-ministerial Invasive Species Council, was effectively built around this figure.

Pimentel *et al.* (2002) have since extended their approach to estimate a global annual loss of $1.5 trillion. These estimates were formed by summing cost data from a series of case studies identified in the IAS literature which were then scaled up to national and international levels using multipliers. As costs include both losses and costs of control, and it is not clear if the aggregate figure represents the impacts of IAS *per se* or responses to them.

While these studies have been important in raising awareness of the potential magnitude of the IAS problem, they do not provide a solid platform for the development of IAS-specific policy when it is needed. Species introductions today may cause economic, environmental and social problems tomorrow that are different from those anticipated by our current markets and government agencies. So, species or pathway-specific actions somehow need to be informed by forward-looking, multi-faceted decision support tools rather than backward looking, aggregated assessments.

The externality problem with IAS will remain relevant for the foreseeable future. For as long as agriculture has existed insects, plants and pathogens have caused outbreaks that have reduced crop and livestock production in and between regions. Industry self-regulatory and government institutions have emerged to prevent introductions of IAS, yet agricultural losses to IAS have remained considerable.

These losses not only include physical production losses, increased growing costs and prevention of trade, but also include environmental and social losses. Although more difficult to tally than agricultural losses, ecologists have long warned of severe environmental consequences of IAS (Williamson, 1996). This view is reflected in international agreement, embodied in the Convention on Biological Diversity (1991), that countries should prevent, eradicate or control species which threaten local species, habitats or ecosystems. This gives recognition to the dual effects of environmental bioinvasions — reduction of native biodiversity (including the extinction of native species), and the disruption of ecosystem services.

While agricultural and environmental systems both face growing threats from IAS, non-agricultural risks tend to receive less attention due to a lack of credible quantified evidence. For example, following a Canadian request to access Australia's salmon market in 1994, Australian quarantine authorities commissioned the Australian Bureau of Agricultural and Resource Economics to prepare an economic analysis indicating potential economic damages that could result from diseases considered an importation risk (McKelvie *et al.*, 1994). No environmental risks were considered. Instead, the threat of sizeable damage to commercial fisheries that might be caused by the diseases Furunculosis and Infectious Haematopoietic Necrosis was used as the basis for the Australian government to refuse Canada's request, which prompted Canada to take the matter to the WTO's Dispute Settlement Body. The Panel failed to uphold Australia's salmon import ban (WTO, 2000), a decision which was appealed by the Australian government on behalf of its aquaculture industries.

Throughout this long-running case, Australia continued to import aquarium fish and herring bait, which carry far greater disease risks than salmon to both commercial fisheries and native marine and estuarine ecosystems (Cook *et al.*, 2011b). These risks were not subjected to the same amount of scrutiny, and did not garner the same sort of political and social attention in Australia as the risks to commercial fisheries. Hence, the government intervention had done little to affect the actual disease risk. This is another example of how governments fail when dealing with IAS, just as markets do.

While some international risk assessment processes incorporate environmental and social risks associated with IAS (e.g. Baker *et al.*, 2008; Pheloung *et al.*, 1999), a lack of quantitative information about their extent can mean deferral to other decision criteria. Environmental and social assets do not have the same easily expressed annual values as agricultural assets. It follows that a quantitative economic criterion tends to carry greater weight in terms of influence over strategic investments in IAS risk-mitigation since it can be readily informed with data.

1.3 Objectives of the Book

To date the vast majority of research into the impacts of and policy responses to IAS outbreaks has been journal-based, with the remainder typically in the form of government research reports. In addition, due to funding and data problems, few researchers have had the opportunity to specialise in the analysis of biosecurity policy, resulting in a very small field of researchers with the expertise for providing such analyses.

Given this, there is considerable demand for improved understanding of the evaluation of biosecurity policy, but a shortage of research capability for doing so. This book is intended to aid biosecurity policy decision makers, as well as advanced students and other academic researchers in this area. Its value to readers can be understood from a more detailed description of its contents. Specifically, this brief introduction to the issues confronting plant biosecurity policymakers and how IAS risks have been assessed in the past is followed by the development of the ecological model (Chapter 2) and its integration into a bioeconomic model to create the IAS impact evaluation framework (Chapter 3). Subsequently, a broad range of case studies is presented to demonstrate the applicability of the interdisciplinary methodology developed previously in the book (Chapter 4). Based on the findings of these case studies the book then reviews some of the particular complexities in undertaking such evaluations:

(1) The varying patterns of IAS impact (Chapter 5);

(2) The choice between prevention and eradication policies towards IAS outbreaks (Chapter 6);

(3) The problem of environmental and social impacts often being non-monetised (Chapter 7).

The book ends with a Concluding chapter which considers a series of questions for future biosecurity policy (Chapter 8).

CHAPTER 2

AN ECOLOGICAL MODEL FOR PLANT PEST AND DISEASE OUTBREAKS

2.1 Introduction

In this chapter we put forward a generic biological model that can be used to predict the effects of invasive alien species (IAS) when they are introduced to new ecosystems.

For decision-support analysts, predicting the risks associated with exotic or non-indigenous IAS requires probability information about each step in the invasion process, including entry, establishment, spread, and impact creation (Cook *et al.*, 2007; Biosecurity Australia, 2006). Yet, for many organisms, we know very little about these steps, and more often than not lack a quantitative knowledge of them.

Even if we do have quantitative information from one or more observed invasion events, we probably still have insufficient information to be able to predict invasion consequences. Even under strictly controlled experimental conditions, endogenously generated variance in spread rate can be high. (Melbourne and Hastings, 2009). So, there are inherent limits to predictability.

This partially explains why quantitative IAS risk assessments are rare (Andersen *et al.*, 2004; Bossenbroek *et al.*, 2005). Most risk assessment protocols, such as the widely adopted weed risk assessment in Australia (Gordon *et al.*, 2008; Pheloung *et al.*, 1999), are based on expert opinion and qualitative assessment, rather than rigorous quantitative statistics. Quantitative approaches for IAS risk assessment have been developed (e.g. Kolar and Lodge, 2002), but are exceptions rather than a norm.

While a lot of progress has been made in developing risk assessment for IAS, the assessment alone will not necessarily enhance predictability (Crowl *et al.*, 2008; Pyšek and Richardson, 2010). Only a small proportion of introduced species become invaders (Pyšek and Richardson, 2010). The chance for an imported plant becoming a weed in Australia, for instance, ranges from 0.007% to 17%, with a central tendency of 2% (Smith *et al.*, 1999). This low probability means there are relatively few data points with which to study biological invasions and any existing information may not be representative (Franklin *et al.*, 2008). Additionally, IAS researchers and decision-support analysts tend to work on IAS with imminent or realised impacts because of funding availability, in turn generated by political impetus (Pysek *et al.*, 2008).

For our purposes and for those of policy support analysts more generally, decision making requires tools that are explicit about the uncertainty in invasion processes. Moreover, these tools should be able to help us to generate predictions despite us lacking data on which to base them. Fortunately, we can draw on a rich history of predictive ecological modelling to help us.

2.2 A Generic Spread Model

While detailed models have been developed to predict the dynamics of spread for some IAS, parameterisation of these models commonly requires large investment in ecological research and, consequently, they are not feasible for most species. Our approach, rather, is to develop a generic model which captures the key ecological processes involved in the spread of most invasive organisms, and which is able to reliably simulate the invasion process using parameters that are either easily measured empirically, or estimated using knowledge of the biological classification and/or ecology of the IAS concerned.

We do not necessarily advocate the use of the model we derive here in every situation. In the case studies to follow, we require a consistent modelling approach to make comparisons, but in other circumstances analyses might involve more specific models. So long as the model is rigorous and justifiable, we see no reason why it cannot be used. Indeed, if time and money were not scarce it would be ideal to employ multiple models to estimate potential IAS spread over time.

Our spread model is based on a conceptual model identifying the three principle processes in the biology of IAS invasions (Liebhold and Tobin, 2008): arrival (the process of transportation of individuals or propagules to new areas outside the native range), establishment (population growth to form a self-sustaining population); local population growth and spread (the increase in population size and associated expansion of the area occupied); and satellite generation (Figure 2.1). We differ slightly from other attempts to generalise the invasion process (e.g. Liebhold and Tobin, 2008; Hastings *et al.*, 2005) in that we separate the radial spread of established populations, caused by population growth and normal dispersal of individuals, form the process of satellite population initiation; processes that are commonly grouped within a single spread function. This is appropriate because long-range IAS dispersal is often associated with human-mediated transport.

The conceptual framework of Figure 2.1 forms the basis of a simulation model for species invasions and impact that we will use later on. The challenges in developing this model were to capture the general ecological processes of species invasion, while maintaining relevance to a broad a range of types of invasive organisms. In Chapters 3 and 4 we will integrate this model with an economic model to predict the economic impact over time of future IAS introductions.

Moving from the conceptual model of Figure 2.1 to a formal model we can apply to case studies, let us denote areas or regions in which host plants are located as i and the probability of an IAS arriving in that region as z_i. If the IAS is present in a region in the first time step of the model, $z_i = 1$. If it is not present, the transition from a *without* to a *with*-IAS situation must be modelled

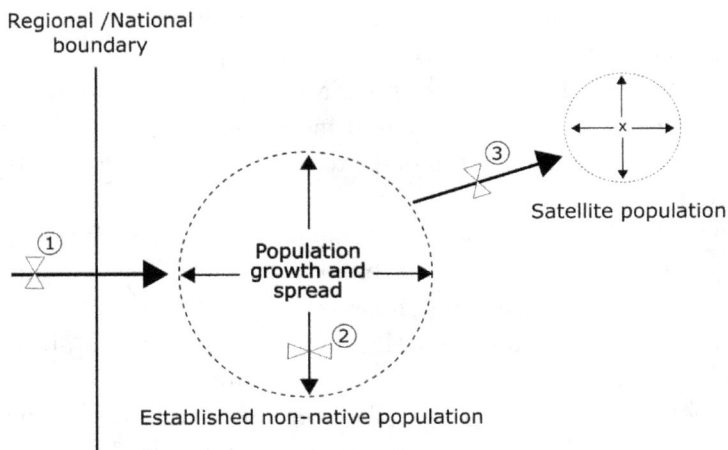

Figure 2.1. A conceptual model of species invasion

Note: The main processes are identified: (1) Arrival and establishment, which is mediated by large scale dispersal constraints caused by geographical factors such as land bridges, air and ocean currents as well as anthropogenic factors such as accidental or deliberate movement of the organisms or their hosts; (2) Local population growth and spread, which is mediated by abiotic (climate, geology, biogeochemistry) and biotic (predators, parasites, hosts, competitors and mutualists) constraints, which in turn may be modified by human activity (e.g. land-use); and (3) Satellite generation, which is mediated by natural dispersal constraints, landscape connectivity and anthropogenic movement on species or their hosts.

in subsequent time steps. This transition is represented in our model as a Markov process.

Consider the probability of an IAS arriving in a region (call it event x) in a time step, $t+1$, conditional on its absence (call it event y) in time step t, as z_{xy}. There is also a probability attached to event x occurring in both time steps, denoted z_{xx}. All possible outcomes for time step $t+1$ are arranged in a transition matrix, $Z = \left(\begin{smallmatrix} z_{xx} & z_{xy} \\ z_{yx} & z_{yy} \end{smallmatrix} \right)$. The elements in the matrix are conditional probabilities indicating the probability of experiencing event x in a time step given that a certain state (i.e. either y or x) was observed in the previous time step.

The methodology of Waage *et al.* (2005) is used to estimate each of the transition probabilities using uniform (or rectangular) distributions taken from Biosecurity Australia (2001) to represent the probabilities of entry (z_{xy}) and establishment (z_{xx}). The remaining transitional probabilities are calculated as $z_{yx} = (1 - z_{xx})$ and $z_{yy} = (1 - z_{xy})$.

Denoting the probabilities of the events y and x occurring in any time step $z_y(t)$ and $z_x(t)$, respectively, and $z(t)$ is a column vector with elements $z_y(t)$ and $z_x(t)$ (where $z(t) = (z_x(t), z_y(t))'$), the transition matrix can be used to calculate the probability of x occurring in time period $t + 1$ as $z_x(t + 1) = Z_z(t)$. The vector $z(t)$ will converge to a unique vector as t increases (Moran, 1984; Hinchy and Fisher, 1991), and the probabilities of event y or x occurring in any given time step reduce to constant values after several time steps. Since event x (i.e. arrival) is of primary concern in a growing region i, $z_x(t)$ is henceforth denoted as z_{it}.

To describe the dispersal of an IAS across multiple regions after its arrival, a stratified diffusion model that combines both short and long distance dispersal processes is used (Hengeveld, 1989a). It is derived from the reaction–diffusion models originally developed by Fisher (1937) and Skellam (1951) which have been shown to provide a reasonable approximation of the spread of a diverse range of organisms (Okubo and Levin, 2002; Dwyer, 1992; Holmes, 1993; McCann *et al.*, 2000; Hastings *et al.*, 2005; Liebhold and Tobin, 2008). These models assert that an organism diffusing from a point source will eventually reach a constant asymptotic radial spread rate of $2\sqrt{r_i D_{ij}}$ in all directions, where r_i describes intrinsic rate of population increase for an IAS per year in region i (assumed constant over all affected sites) (Lewis, 1997; Shigesada and Kawasaki, 1997; Hengeveld, 1989a). D_{ji} is a diffusion coefficient for an infested/infected site with an age index j in region i, and is also assumed constant over time for simplicity. Hence, assume that the original outbreak takes place in a homogenous environment in region i and expands by a diffusive process such that area affected at time t, a_{ijt}, can be predicted by

$$a_{ijt} = z_{it}(4D_{ij}\pi r_i t^2). \tag{2.1}$$

The density of an IAS within a_{ijt} influences the control measures required to counter the effects of an infestation/infection. Assume that in each site with age index j in region i affected, the density (N_{ijt}) of the organism grows over time following a logistic growth curve until the carrying capacity of the environment (K_{ij}) is reached:

$$N_{ijt} = \frac{K_{ij} N_{ij}^{\min} e^{\delta_i t}}{K_{ij} + N_{ij}^{\min}(e^{\delta_i t} - 1)}. \quad (2.2)$$

Here, $N_{ij}^{\min}$ is the size of the original influx at a site with age index j in region i, and δ_i is the rate of density increase in region i. In the case studies to follow we assume $r_i = \delta_i$.

The area occupied by the IAS is also influenced strongly by the initiation on new populations, or *satellite* infestation sites as referred to by Moody and Mack (1988). The number of satellite sites in year t, s_{it}, can take on a maximum value of $s_i^{\max}$ in any year. Satellites result from events external to the outbreak itself, such as weather phenomena or human behaviour, which periodically jump the expanding infestation/infection beyond the invasion front (Cook et al., 2011a). We use a logistic equation to generate changes in s_{it} as an outbreak continues

$$s_{it} = \frac{s_i^{\max} s_i^{\min} e^{\mu_i t}}{s_i^{\max} + s_i^{\min}(e^{\mu_i t} - 1)}. \quad (2.3)$$

Here, μ_i is the rate of new foci generation in region i, and $s_i^{\min}$ is the minimum number of satellite sites generated in region i.

Given the predicted area affected by an IAS in sites that have been affected for different lengths of time (i.e. Equation (2.1)), the density of these populations (Equation (2.2)) and the number of satellite sites predicted to have been created (Equation (2.3)), the total area, A_{it}, is calculated across m sites as

$$A_{it} = s_{it} \sum_{j=1}^{m} (a_{ijt} N_{ijt}), \quad \text{where } 0 \le A_{it} \le A_i^{\max}. \quad (2.4)$$

Here, $A_i^{\max}$ is the total area planted to hosts in region i.

2.3 Behaviour and Validity of the Biological Spread Model

2.3.1 *Patterns of IAS introduction and establishment*

Insect pests of crops have a long history of international spread, and like plant diseases, national risk assessment and quarantine procedures for such species are well developed. The International Plant Protection Convention (IPPC) and regional plant protection organisations, like the European Plant Protection Organization (EPPO), provide global support to the development of protocols and sharing of information.

Most new introductions of terrestrial invertebrates are accidental, involving transport of undetected individuals on agricultural materials (e.g. in soil or as eggs on plants) or other objects (e.g. as insect larvae in wood used for packing, as free insects in aircraft cabins). It has been suggested that burgeoning international trade has driven species invasions, evidenced by strong temporal correlations between the establishment of IAS and volume of imports (Levine and D'Antonio, 2003) and global patterns of species exchange which reflect global trade routes (Van Kleunen *et al.*, 2015), though causation is often far from clear (see Section 2.4).

Establishment and spread of invertebrates is particularly sensitive to abiotic factors such as temperature for continental Europe but spreads to UK only through produce, and may not thrive in the UK climate, if ever established (Baker *et al.*, 2000), though future climate change may allow establishment of such species. Together, increasing trade volumes, growth of new trade routes and changing climate are likely to increase the rate of colonisation of IAS, and historical data support this prediction (Waage *et al.*, 2005).

2.3.2 *Validity of the spread model*

The constant rate of increase in radial range predicted by Fisher–Skellam type reaction–diffusion (i.e. $2\sqrt{r_i D_{ij}}$), suggests that the square root of occupied area should increase linearly with time (area increases exponentially). There have been extensive studies to test this prediction against recorded spread of IAS. Shigesada and

Kawasaki (1997), for instance, discuss the forms that the relationship between the square root of occupied area and time can take with reference to past IAS invasions: Type 1 – a linear increase; Type 2 – biphasic increase with an initial slow slope followed by a steep linear slope; Type 3 – an accelerating non-linear curve with the gradient increasing with time. By separating radial spread from the generation of satellite populations, our model allows simulation of Type 1 curves ($\mu = 0$) and Type 3 expansions ($\mu > 0$). We do not explicitly simulate Type 2 curves though we do not see this as a major drawback since Types 2 and 3 patterns are often difficult to distinguish with real data (Shigesada and Kawasaki, 1997), and both are postulated to result from satellite generation.

General patterns can be observed in empirical studies of range expansion by IAS. Firstly, many species of mammal, bird, insect and plant do exhibit Type 1 dynamics suggesting that the Fisher–Skellam model is an appropriate framework (Shigesada and Kawasaki, 1997). Agassiz (1994) reviewed the spread of 25 species of Lepidoptera (butterflies and moths) that invaded Britain in the last century and found that 21 species exhibited Type 1 spread dynamics. Secondly, Type 2 or 3 dynamics are sometimes exhibited by birds (Mundinger and Hope, 1982; Hengeveld, 1989b), insects (Andow *et al.*, 1993) and plants (Moody and Mack, 1988) and this is usually attributed to long distance or 'jump' dispersal, as represented by a non-zero value for the satellite generation parameter μ in our model.

There have been several studies which have aimed to compare spread in observed invasions with the spread rate predicted by the reaction–diffusion model. Generally such validation is done by local measurements of r and D and comparing the model output with larger scale records of the rate of spread. The possibility of independently estimating r and D is a major advantage of this modelling framework: r is the instantaneous rate of population growth that can be derived from measurements of population size over time (i.e. $\frac{\ln(\text{popn size}_{t+1})}{\ln(\text{popn size}_t)}$), or can be estimated from the fecundity and longevity of the species, while D can be estimated as the standard deviation of dispersal distances, typically measured by mark, release, recapture studies (Kareiva, 1983). A good match between predicted and observed spread has been found with a large number of plant-feeding insects, while most failures of prediction are not ascribed to poor parameterisation, but to a failure to account for long-dispersal dispersal (Liebhold and Tobin, 2008), such as in the case of the cereal leaf beetle (Andow *et al.*, 1990), and this is the primary justification for the separation of satellite population generation from radial spread in our model.

2.4 Observed Dynamics of IAS Spread

Terrestrial invertebrates are one of the best-studied groups of IAS, and we have a relatively large data collection of data on their rates and patterns of spread following naturalisation (Table 2.1). Even so, relatively few studies have measured individual population parameters of the invading organism in addition to the overall rate of spread. For many of this group the square root of occupied area increases linearly with time, suggesting that the diffusion model fits the data adequately (satellite creation can be assumed to be zero).

Many of the arthropod IAS exhibiting very rapid spread are agricultural pest species, likely because they are generally highly dispersive and because movement on transported crops or other materials can result in rapid production of satellite populations. Fortunately, most current agricultural outbreaks are local, often in protected habitats where control is relatively easy. Many of the species for which we have data are moths invading more natural habitats, which tend to exhibit slower rates of spread, though the populations are likely to be more difficult to detect and contain.

For plant diseases, data on the rate and pattern of spread are less readily available than for arthropod herbivores. The data that do exist reveal that these organisms can spread very rapidly, sometimes across continental scales in individual years, though it is often difficult to determine whether the organisms are dispersing, or whether the apparent population spread actually represents a wave of local infection through space caused by the emergence from local resting stages in response to climatic or other variables. Experimental tests of disease propagation at small temporal and spatial scales do,

Table 2.1. Parameter values found in the literature for the spread of non-native terrestrial invertebrate species

Order	Common name	Name	r (year)	D (km² yr⁻¹)	Predicted radial expansion (km yr⁻¹)	Observed radial expansion (km yr⁻¹)	A (km yr⁻¹)	Source
Hemiptera	Hemlock woolly adelgid	*Adelges tsugae*				8–13		Evans and Gregoire (2007)
	Beech scale	*Cryptococcus fagisuga*				4–15		Morin et al. (2007)
Coleoptera	Japanese beetle	*Popillia japonica*				5.5, 27.5		Shigesada and Kawasaki (1997), Shigesada et al. (1995)
	Rice water weevil	*Lissorhoptrus oryzophilus*					28	Andow et al. (1993)
	Great spruce bark beetle	*Dendroctonus micans*					15	Fielding et al. (1991)
	Japanese beetle	*Popillia japonica*				5–6		Shigesada et al. (1995)
	Cereal leaf beetle	*Oulema melanopus*	1.6–1.9	0.4	13–127	14.7–170		(Andow et al., 1990)
Diptera	Tetse fly	*Glossina sp.*			2–25			Hargrove (2000)
		Pseudacteon tricuspis				20		Porter et al. (2004)
	Love bug	*Plesia nearctica*				32		Buschman (1976)
Lepidoptera	Small white butterfly	*Pieris rapae*	9–32	2.4–64	9.3–90	15–170		Andow et al. (1993)
	Gypsy moth	*Lymantria dispar*	4.6			3–29		Tobin et al. (2007)
	Horse chestnut	*Cameraria*				17–39		Augustin et al.

Firethorn leaf miner	*Phyllonorycter leucographella*	10.3	Nash et al. (1995)
	Etainia decentella	3.05	Agassiz (1994)
	Stigmella suberivora	2.2	Agassiz (1994)
Fern smut	*Psychoides filicivora*	1.69	Agassiz (1994)
	Caloptilia rufipennella	6.71	Agassiz (1994)
Ruddy streak	*Parocystola acroxantha*	7.33	Agassiz (1994)
	Agrolamprotes micella	0.056	Agassiz (1994)
	Teleiodes alburnella	3.89	Agassiz (1994)
Dingy dowd	*Blastobasis lignea*	3.72	Agassiz (1994)
	Blastobasis decolorella	2.2	Agassiz (1994)
Carnation tortrix	*Cacoecimorpha pronubana*	2.26	Agassiz (1994)
	Ptycholomoides aeriferanus	2.31	Agassiz (1994)
Light brown apple moth	*Epiphyas postvittana*	2.54	Agassiz (1994)
Summer fruit tortrix	*Adoxophyes orana*	1.07	Agassiz (1994)
Orange pine tortrix	*Lozotaeniodes formosanus*	2.65	Agassiz (1994)

(Continued)

Table 2.1. (*Continued*)

Order	Common name	Name	r (year)	D (km^2 yr^{-1})	Predicted radial expansion (km yr^{-1})	Observed radial expansion A (km yr^{-1})	Source
		Acleris abietana				4.12	Agassiz (1994)
		Pammene aurantiana				2.43	Agassiz (1994)
	Fenland pearl	*Phlyctaenia perlucidalis*				4.68	Agassiz (1994)
	Spruce knot-horn	*Dioryctria schuetzeella*				5.36	Agassiz (1994)
	Balsam carpet	*Xanthorhoe biriviata*				1.81	Agassiz (1994)
	White-banded carpet	*Spargania luctuata*				1.24	Agassiz (1994)
	Cypres pug	*Eupithecia phoeniceata*				2.14	Agassiz (1994)
	Feathered beauty	*Peribatodes secundaria*				3.46	Agassiz (1994)
	Varied coronet	*Hadena compta*				2.54	Agassiz (1994)
	Blair's shoulder-knot	*Lithophane leautieri*				3.78	Agassiz (1994)
	Golden plusia	*Polychrysia moneta*				2.99	Agassiz (1994)
		Phyllonorycter platani				8.63	Agassiz (1994)

however, highlight fast spatial dynamics (Table 2.2). There appears to be a distinction in these data between pathogens of woody species and those of herbaceous species, the latter being orders of magnitude faster.

Dispersal mechanisms are highly variable, and this may account for some of the variability observed (e.g. between fungal spores transmitted by water and wind). Aerial dispersal of spores and movement by insect vectors must certainly contribute to the rapid spread of relevant species, but satellite creation by movement of infected plant material is also important, as has been postulated recently in the United States and Europe for the movement of *Phytophthora ramorum* via ericaceous shrubs distributed widely to garden centres.

2.5 Origins and Drivers of IAS Risks

Whereas it is traditional to think of biosecurity risks being associated with accidental contamination of agricultural (animal and plant) imports, our analysis suggests that these imports account for only a fraction of the problem, and possibly a declining one. A large source of risk is deliberate importation of plant and animal species, for the garden trade, the pet trade, new food production systems (e.g. fish, crayfish production). This risk is twofold — species intentionally introduced into controlled habitats may escape into natural ecosystems, and those escaped species may be contaminated with diseases that will affect native or agriculturally important species. When we come to consider prevention and control, it will be clear that intentional introductions offer substantial opportunities for risk reduction, relative to purely accidental introductions.

Another distinctive aspect of the origin of risk is that introduced by evolution and adaptation of introduced species. Across the taxa surveyed are examples of IAS which hybridise with native species or evolve to adapt to local conditions, thereby creating new problems for both agriculture and the environment. These evolutionary effects are even less predictable than those caused by introduction alone. Richardson *et al.* (2000) and Simberloff and Von Holle (1999)

Table 2.2. Parameter values found in the literature for the spread of plant diseases. Here r refers to the intrinsic rate of increase of the infected population of the host species. Since these are agricultural host species, this rate of increase of infection relates readily to spatial extent of infection

Common name	Name	r (year)	D (km^2 yr^{-1})	Predicted radial expansion (km yr^{-1})	Observed radial expansion (km d^{-1})	A (km^2 yr^{-1})	Source
Herbaceous hosts							
Potato blight	*Phytophthora infestans*	58–153					(Zadocks and Shein, 1979)
Yellow rust	*Puccinia striiformis*	36–99					(Zadocks and Shein, 1979)
Tomato mosaic virus		36					(Zadocks and Shein, 1979)
Woody hosts							
Oak wilt disease	*Ceratocystis fagacearum*	0.77					(Zadocks and Shein, 1979)
Leaf rust	*Cronartium fusiforme*	0.4					(Zadocks and Shein, 1979)
Wilt fungus	*Fusarium oxsporum*	0.5					(Zadocks and Shein, 1979)
Root rot fungus	*Phytophthora cinnamomi*	1.54					(Zadocks and Shein, 1979)
Sudden death	*Valsa eugeniae*	0.34					(Zadocks and Shein, 1979)
Tobacco blue mold					13.9 km/d		(Aylor, 2003)
Wheat stem rust					35 km/d		(Aylor, 2003)

illustrate a number of other ways in which IAS interact with each other or native species to accelerate or exacerbate invasions. These include new associations of species which facilitate reproduction, spread or survival of a non-native, such as dispersal of non-native plant seeds by native or introduced birds. Note that these "second order effects" are not incorporated into the ecological model, and hence not in the economic model to follow. This means that impact of non-natives may be underestimated.

Many authors have associated a growing risk from IAS with the rapid growth of international trade. Trade statistics cited include the value of traded goods, which have risen from $192 billion in 1965 to $6.2 trillion in 2000. Upward trend statistics are also seen in commodity imports and container shipments. Often, however, it is hard to link the introduction of particular taxa with such broad pathways, and we must ask whether positive correlations of general trade and increase in introductions are really evidence of causation? Many of our new problems appear associated with pathways other than the bulk import of traded or even agricultural commodities. Speciality importers of pets, plants, certain foodstuffs (e.g. bushmeat), game fish as well as individual travellers bringing such species into the country, may pose a greater risk from many taxa than large-scale commodity movements. For some taxa, like plants, tomorrow's IAS problems are already here, and the rate limiting process is not introduction but establishment and spread in nature.

A much more refined analysis of changes in trade, travel, transport and tourism will be needed to firmly demonstrate the link between IAS risks and recent global trade liberalisation. However, all institutions consulted for this study have observed that the complete removal of trade restrictions within Europe will have major implications for risk. Recent UK biosecurity crises have shown the ease with which animal (e.g. foot-and-mouth disease (FMD), bovine spongiform encephalopathy (BSE)) and plant (e.g. potato ring rot, sudden oak death) diseases may move between UK and continental Europe via trade, and pan-European movement of new environmental problems (e.g. oak knopper gall, horse chestnut leaf miner), probably by other routes, is now a regular phenomenon.

With Europe's new members, and even more distance and permeable borders, IAS risk may increase substantially.

2.6 Nature and Impact of IAS

There are three important ways in which IAS can affect the economy and society. Firstly, IAS may directly affect production systems, like agriculture, as pests, diseases and weeds. Secondly, due to international biosecurity and trade structures, the appearance of a new IAS may affect trade and the economy, even without having any direct harmful effect. This phenomenon is seen with animal, fish and plant diseases, but may grow in future.

Thirdly, all taxa of IAS pose some kind of risk to the environment. Indeed, this is the most rapidly emerging kind of impact amongst all these taxa. Environmental impact of IAS can be of two kinds:

(1) Non-native species can directly affect biodiversity, reducing the abundance of native species through predation, competition (including the introduction of diseases which have more impact on natives) or parasitism (in the case of introduced parasites or pathogens). Extinction of native species may be a consequence of IAS, but the loss of native plants and animals through hybridisation with IAS is likely;

(2) Non-native species can affect ecosystem processes and services, such as the supply of clean water and air, or the functioning of ecosystems to provide resources which support animal and plant communities in food chains, or ecological succession. Obvious examples include IAS which affect the physical environment, such as the structure of waterways and channels, or the turbidity and quality of water. Invasive plants may affect ground cover and soil structure, while non-native herbivores may do the same by removing vegetation.

While impact on biodiversity is likely to be more obvious and more press-worthy, the impact of IAS on ecosystem services are likely to be more severe in the long term, more cryptic and slower to develop. Domesticated animals and crops/garden plants provide both pathways and reservoirs for diseases of environmental importance, thereby linking agricultural and environmental risks across a number of taxa.

A second aspect of impact involves the rate of spread of IAS. Tables 2.1 and 2.2 highlight the variability of rates of spread both within and between organism classification groups. Not surprisingly diseases emerge as rapidly spreading non-natives, while arthropods may spread more slowly. However, there is a surprisingly large range for some taxa, which cautions against generalisation. The other important feature is the potential for long lag-times in the emergence of IAS problems. Policy decisions about where to focus prevention and eradication will be influenced strongly by the relevant timescale envisioned by policymakers (Cook *et al.*, 2011b).

2.7 Conclusion

In this chapter we have presented a model that can be generally applied to IAS in the absence of data to make predictions about their abundance and distribution over time. This model is a suggested approach rather than one we would advocate in every situation. Its strength is that it is grounded in ecological theory, and hence the parameters on which it relies to describe IAS populations over time are familiar to scientists. That is, they are not abstract constructs to fit the workings of the model, but are instead parameters that we have a certain amount of reliable information about. With these parameters we can at least make informed predictions about possible future IAS incursions bearing in mind that the future is extremely uncertain.

So far, our model only gives us the area of hosts affected over time. In the next chapter, this information is to be combined with tangible impact information. As we will see, this is most appropriate when the hosts are market goods with easily expressed annual values, such as agricultural goods. By combining information about increased costs and decreased revenues resulting from IAS, we can predict the total impact of individual species on social welfare over time; or, at least that component of social welfare indicated by cost and revenue information. It also means we can evaluate policies that could be enacted in response to IAS incursions in terms of likely reductions in future impact.

CHAPTER 3

A BIOECONOMIC MODEL FOR ANALYSING IAS IMPACTS

3.1 Introduction

Chapter 2 has given us an ecological foundation with which we can estimate the area of agricultural hosts affected by a wide range of IAS. As mentioned, this may not be best spread model to use in every situation, but it is sufficiently flexible to represent a wide range of species. In this chapter, we will discuss how a spread model can be combined with economic parameters to calculate the costs of agricultural IAS over time. As the agricultural assets included are privately owned, the values we concern ourselves with are market-based impacts. We first present some of the concepts we will use to estimate the agricultural costs of IAS before moving to a model that can be practically applied to IAS policy decisions.

3.2 Conceptual Model

3.2.1 *Specification*

A biological model such as the one presented in Chapter 2 allows us to predict the likely spread of an IAS when it is introduced to a new environment, but what we really need is an economic model that converts this area to a cost that can be presented to policymakers. Monetary values enable easy comparisons with other investments so that IAS can be assimilated into policymaking processes, be they government or private. We will discuss in Chapter 7 instances where this is not possible, particularly where environmental and social asset damage occurs, but for now we focus on agricultural or industrial damage resulting from IAS and how we can estimate this based on changing abundance over time.

To do so, we need to make several economic assumptions. Economists are often criticised for assuming away complications, but it is far better to make these explicit so that results can be properly contextualised. For the moment, let us assume there is one IAS of concern to a particular country or region, and that this organism has an impact on one known agricultural good grown in a homogenous environment. Secondly, assume the domestic market for the host commodity is perfectly competitive, implying product homogeneity. Thirdly, assume that the contribution of domestic producers of that affected commodity to the total world supply is insufficient to exert influence on the world price, the exchange rate and domestic markets for other commodities.

Box 4. What is a Perfectly Competitive Market?

A perfectly competitive market is one characterised by a large number of profit-maximising producers supplying a homogenous good. There are no impediments to new suppliers entering this market, nor penalties for those wishing to leave it. Each supplier to the market is a price-taker, meaning that the price the supplier receives for output is dictated by the market. The supplier only makes decisions about the quantity it will produce and the amount of inputs purchased to produce that output. Perfectly competitive markets are rarely observed in the real world with the exception of agricultural markets.

A static, partial equilibrium model can be used to examine the economic implications of IAS. Consider a grower producing an agricultural product q. The grower's production function describes the relationship between physical quantities of factor inputs and the physical quantities of output involved in producing q given the state of technological knowledge possessed by the grower. The inputs, denoted (I), include both fixed (e.g. buildings and machinery) and variable (e.g. fertiliser, chemicals, fuel) components. The level of

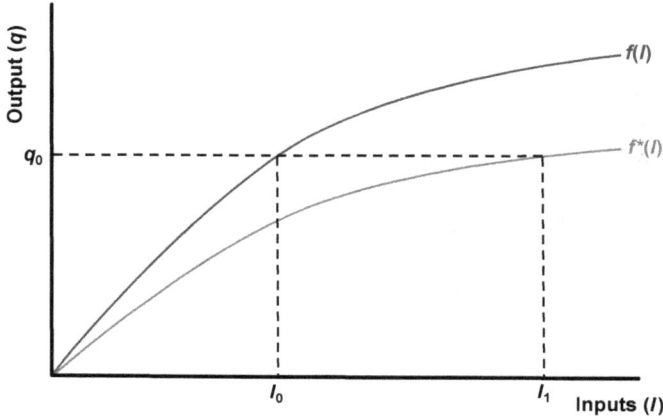

Figure 3.1. The production function with and without an IAS

output produced by the grower is some function, call it f, of I:

$$q = f(I). \tag{3.1}$$

For the moment, assume any uncertain factors in the production process take on their average values.

Figure 3.1 provides a graphical representation of possible production functions with and without an IAS, labelled as $f^*(I)$ and $f(I)$, respectively. Generally, to be of biosecurity significance an IAS has a negative impact on output when established in a production area (Waage *et al.*, 2005). In this example, the grower must purchase I_0 to produce quantity q_0 under normal circumstances, but when an IAS is introduced requiring additional chemical applications the necessary inputs increase to I_1. Thus, an IAS impact can be seen in much the same light as a negative technological change.

To examine the welfare implications of IAS-induced change requires some discussion about the producer's cost and revenue functions. In short, the total revenue (TR) earned by the producer depends on the quantity sold and the price (p). As the producer is a price-taker, p is dictated to her by the market. Her TR is given by

$$\text{TR} = pq. \tag{3.2}$$

Box 5. General Equilibrium Analysis

General equilibrium analysis provides a broad indication of how changes in
the likelihood of an adverse shock in one sector of the economy, such as
an IAS incursion in a large national agricultural industry, could affect the
entire economy. It does so by capturing important production processes
through which flow-on effects to the rest of the economy can be felt, and
how these in turn might feed back into the original sector affected. General
equilibrium analysis therefore relies on describing the nature and strength
of inter-sectoral linkages using techniques such as input–output analysis.
This approach is preferable to partial equilibrium analysis where a sector
with substantial linkages to the rest of the economy is likely to be affected.
However, in cases where IAS affect smaller intensive agricultural industries
the broader economy may not experience measurable change.

Total costs are a function (call it c) of output.

$$\text{TC} = c(q). \tag{3.3}$$

The production function of Figure 3.1 tells us that different levels
of output require different purchases of inputs, but that increases in
output achieved with successive increases in inputs are not constant.
This occurs regardless of the type of input and hints at the shape
of the function c. If a grower were only to vary her fertiliser
input, for example, while holding all other inputs constant, she
would achieve smaller incremental (or 'marginal') output gains with
successive fertiliser applications. At first, the marginal output gains
might increase if there are sub-optimal nutrient levels in the soil to
begin with, but gradually plants will benefit less, and eventually be
adversely affected by additional fertiliser applications.

In Figure 3.2 we have plotted the grower's TR and TC functions.
As her output does not have an influence on price, the TR function
is represented as a straight line from the origin. The slope of this
line represents the price she receives, and also represents marginal
revenue (MR). Similarly, the slope of the TC function represents
marginal cost (MC), but unlike MR it is not constant. The vertical
distance between TR and TC is maximised where MR = MC. As we

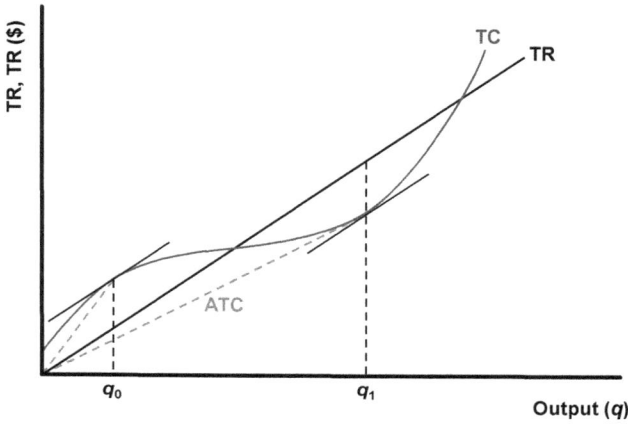

Figure 3.2. Total cost, TR and profit-maximisation

have constructed Figure 3.2, this occurs at two levels of output, q_0 and q_1. However, at q_0 the TC function is above the TR function, indicating that the total returns to the grower are negative at this level of output. At q_1, MR exceeds MC so returns are both positive and maximised at this level of output. These returns represent grower profit, π.

In our analysis, we can think of profit as a proxy for a grower's *welfare*. Welfare is an ambiguous concept, and in reality grower welfare will almost certainly be derived from an array of economic, social and environmental determinants. We will focus on the specific component of welfare determined by π, and the more of it she has the higher the level of welfare enjoyed by the grower. Her motivation for profit-maximisation can be stated as:

$$\max_{q} \pi = pq - c(q). \tag{3.4}$$

To simplify our discussion $c(q)$ will not be divided into its fixed and variable components, as above. Hence, assume fixed costs of production are negligible, so total costs of production are the same as variable costs. This being the case, note that profits are maximised for the grower represented in Figure 3.2 at the point where her average variable costs (AVC) are minimised. AVC are shown for q_0

and q_1 as straight lines from the origin to the points on the TC curve corresponding to each level of output. At q_1, AVC are at their lowest.

Box 6. Profit-Maximisation

The assumption we have made regarding the producer's profit-maximisation motivation ignores other potentially powerful welfare determinants. If they are solely motivated by the goods and services they can purchase, then profit-maximisation would probably be the producer's underlying 'reason for being'. However, less tangible factors such as altruism, respect and influence might also play a role in welfare determination and are not necessarily maximised via the pursuit of profit. Indeed, the connectedness, size or innovativeness of a grower's operation might be better performance indicators to target for these types of welfare determinants. Nevertheless, the assumption of profit-maximisation makes life that little bit easier for analysts trying to model producer behaviour such that they respond to shocks in predictable ways. A model, after all, is a simplified version of reality that can be useful in helping our understanding of a problem, rather than a complete replication of it.

It is not necessarily the case that the producer's choice of output of q will be greater than zero. Where the minimum value of AVC exceeds the prevailing market price it is in the interests of a profit-maximising grower to produce no output in order to minimise her losses. If this were the case the TC curve of Figure 3.2 would lie everywhere above the TR function.

It is now straightforward to derive the supply function for the grower. The supply function tells us how much output she will supply to the market at different prices. At prices above the minimum value of AVC, we have seen that she maximises π at the level of output corresponding to the point where $\text{MC} = \text{MR}(=p)$. It follows that the MC function relates her profit-maximising output to price, and thus represents her supply curve, $q(p)$.

The supply curve for a collective industry is formed by the horizontal summation of all the supply curves of growers producing output for the market. If there are n suppliers and the supply curve for the ith farm is denoted $q_i(p)$, then the supply curve for the industry (S) is given by

$$S = \sum_{i=1}^{n} q_i(p). \tag{3.5}$$

The industry supply schedule S formalises the relationship between industry output and collective MCs of production, and can be used to calculate industry profit under different production conditions.

Returning now to the with- and without-IAS production functions of Figure 3.1, the implications of an IAS for a grower's profit-maximising output decision become clear. As the level of inputs needed to produce each unit of output increases due to costly efforts to mitigate the on-farm impacts of the IAS, so too must MC and AVC.

Exploring the implications of this from an industry perspective, let us first consider the situation where growers are unable to supply a product as cheaply as overseas suppliers. This means that the domestic price of the good will be above an international (or world) price and domestic growers will consequently focus on local supply rather than export. Identical imported products can be sourced by consumers at the world price, but this raises the probability of an IAS being inadvertently imported as a contaminant. To protect against this biosecurity authorities may impose phytosanitary measures that reduce IAS risk, but they are not guaranteed to be 100% effective.

3.2.2 *An importing host industry*

We can demonstrate the impact of an IAS becoming established in an importing market situation like this using a partial equilibrium framework. It is *partial* in the sense that it only depicts one market — the market for the imported host product — as opposed to the broader economy. In other words, we do not consider the impacts on farm service or supply industries, or value-added activities that use the affected product as an input. We just consider the host industry itself and the growers that make it.

Recalling the characteristics of $c(q)$, the AVC curve of each grower will be U-shaped, as depicted in the left frame of Figure 3.3. Here, two sets of cost curves are shown dealing with a with- (MC* and AVC*) and without-IAS (MC and AVC) scenario.

We have established that a profit-maximising grower in a perfectly competitive market will choose to produce a level of output corresponding to the point where price equals the MC of production, which maximises the difference between TC and TR. Assuming the world price, p_w, is below the domestic (or closed) market equilibrium price (i.e. p_d in the right-hand frame of the diagram), a grower characterised by the cost curves MC and AVC would choose to produce quantity q_0 (i.e. where $p = MC$) and earn a profit of $ABCp_w$ in the absence of an IAS. Once again, note that output will be positive as long as the price received by the grower remains above the minimum value of her AVC in the left-hand frame of the figure.

If all growers in the industry behave similarly, the industry supply schedule produced by the horizontal summation of each producer's output at different prices would resemble the curve S in the right-hand frame of Figure 3.3. According to the industry

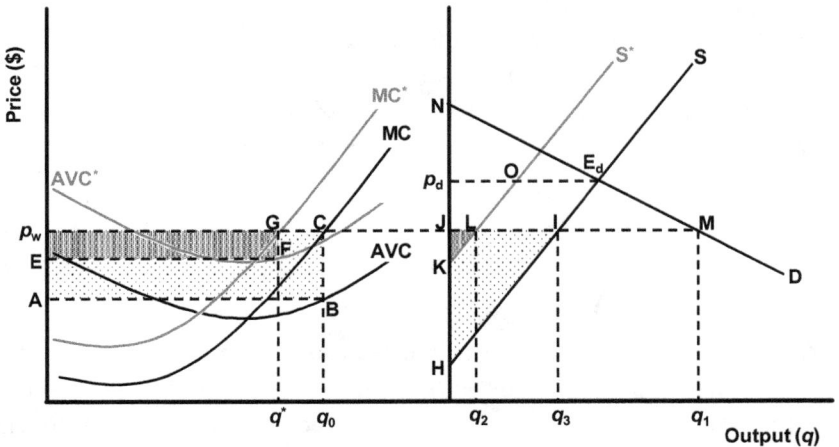

Figure 3.3. The economic impact of an IAS — imported goods in a perfectly competitive market

demand schedule (D), domestic consumers will demand the quantity q_1 at price p_w. Of this, q_0 will be supplied by domestic growers, and $q_1 - q_0$ by imports, assuming there are minimal impediments to trade (such as phytosanitary measures imposed on imported goods).

Losses to the industry resulting from an IAS incursion can be estimated in the partial equilibrium model as the total expected change in *producer surplus* brought about by the induced negative supply shift. Producer surplus is net revenue earned from the sale of a good at prices above the minimum acceptable price growers would have been willing to sell for before having to leave the market. It represents the collective profits of growers within the industry minus fixed costs of production, which again we have assumed are minimal.

The welfare of consumers can also be calculated as *consumer surplus*. Consumer surplus is analogous to producer surplus. It is the financial equivalent of the extra utility gained by consumers from purchasing a good at a price lower than what they were willing to pay for it. Consumer surplus arguments are often excluded from biosecurity welfare analyses in Australia. We will discuss this in more detail in Chapter 7.

In the right-hand frame of Figure 3.3, producer surplus is represented by the area below the prevailing market price line (i.e. p_w) and above the supply curve (i.e. S* with and S without the IAS). Consumer surplus is represented by the area above the price line and below the demand curve. As we have constructed the figure, producer surplus is the shaded area HIJ, and consumer surplus is the shaded area JMN. Note that under a domestic closed economy equilibrium scenario (i.e. E_d) producer surplus would be the larger area $p_d E_d H$, and consumer surplus the smaller area $p_d E_d N$. Hence, the gains from trade are shown as $E_d MI$.

If an IAS were to now enter the production region and become established, the effect at the farm level will be rising AVC (and MC). A greater cost is now involved in producing each unit of output after the outbreak than before it. At the market price p_w the increased costs of production would lower producer output from q_0 to q^* where producer surplus is the heavily shaded area $EFGp_w$.

Box 7. Why are Consumer Surplus Arguments not Considered?

Biosecurity decisions made at all levels of government have the potential to be challenged by World Trade Organization Members. The *Agreement on the Application of Sanitary and Phytosanitary Measures* (known as the SPS Agreement) requires central government signatories to support the observance of the international provisions by all tiers of government (GATT, 1994). This means that regional IAS-freedom statuses protected through the use of phytosanitary measures must be accompanied by an assessment of potential damage to domestic industries being prevented, and that this assessment be consistent with the provisions of the SPS Agreement. In its current form, the SPS Agreement (Paragraph 3, Article 5) stipulates that any welfare effects be measured in terms of producer welfare, while consumer gains from trade are not mentioned. Hence, a region like WA is obliged to consider the risks of IAS to its domestic industries, but is not obliged to consider consumer welfare; much to the chagrin of trade economists and consumer welfare advocates.

If the probability of the IAS entering and becoming established is z, then the expected loss of producer surplus at the farm level is $z \times (\mathrm{ABC}p_{\mathrm{w}} - \mathrm{EFG}p_{\mathrm{w}})$. At an industry level, the domestic supply curve will contract from S to S* in the right frame of Figure 3.3 in the face of added growing costs. Domestic producer surplus will decline to the heavily shaded area KLJ. This represents a loss of HILK, meaning the expected damage to the industry from the IAS is $z \times$ HILK.

To these costs must be added any expenses associated with government or industry response efforts to try to eradicate or contain the IAS. We will discuss this response further below along with policy arrangements currently dictating the form it takes in the Australian context, but for now we simply denote these costs RC.

Using the right-hand frame of Figure 3.3, the expected total damage cost to import-competing domestic producers (C_z^I) with an IAS arrival probability z is given by

$$\mathrm{C}_z^I = z \left[\left(p_{\mathrm{w}} - \int_0^{q_3} S \right) \cdot dq - \left(p_{\mathrm{w}} - \int_0^{q_2} S^* \right) \cdot dq + \mathrm{RC} \right]. \quad (3.6)$$

Box 8. The Effects of Phytosanitary Measures Placed on Imports

If phytosanitary measures are imposed on goods imported from foreign suppliers to lower the transfer risk of IAS, the world price will effectively rise according to treatment costs. This will have the effect of lowering the consumer surplus achieved through trade, but raising producer surplus as growers are effectively cushioned from foreign competition. This is precisely why phytosanitary measures are viewed as non-tariff trade barriers. They are not tariffs, but they work in the same way by restricting international or inter-regional competition. In some cases, the costs of phytosanitary measures may be high enough to prevent non-domestic producers from supplying any product to the market at all. For instance, if phytosanitary measures were to be introduced to goods imported to the market depicted in Figure 3.3 that increased the landed price from p_w to p_d, producer surplus would be $p_d E_d H$ without the IAS and $p_d OK$ with it. Consumer surplus would be $p_d E_d N$ with the phytosanitary measures in place, rather than the larger area JMN without them.

Equation (3.6) states that given a probability of IAS arrival z, C_z^I is equal to the expected difference between the producer surplus if the volume of imports is positive and no outbreak occurs and the producer surplus if an outbreak occurs plus the expected cost of incursion response.

3.2.3 *An exporting host industry*

Now consider a situation in which growers are able to produce a crop relatively cheaply and sell it to other countries at a lower price than they can produce it themselves. This is illustrated in Figure 3.4. Here, the domestic demand curve is once again shown as D in the right-hand frame of the figure, and the pre-IAS industry supply curve is shown as S_0. At the world price p_w, the domestic demand schedule reveals the industry is willing to supply q_4, while the domestic demand is only q_3. The industry can sell the residual $q_0 - q_1$ and earn a total producer surplus of ABC. Consumer surplus is the area MNC. A producer within the industry

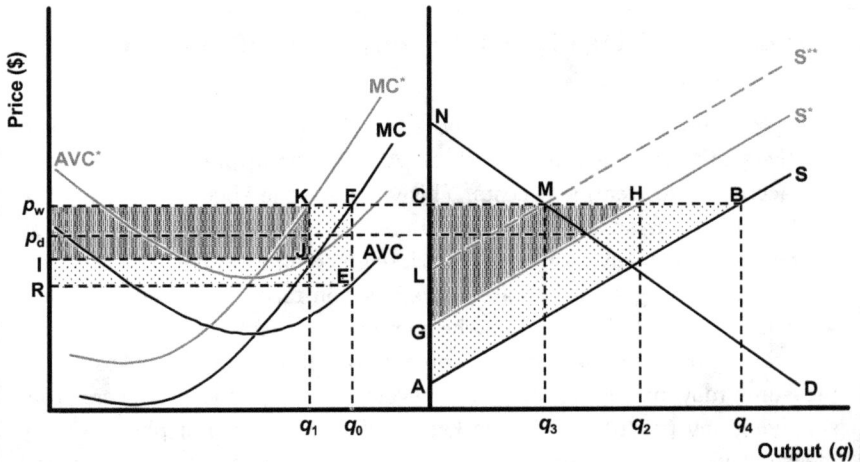

Figure 3.4. The economic impact of an IAS — exported goods in a perfectly competitive market

characterised by the cost curves AVC and MC in the left frame of the diagram earns a profit of $REFp_w$ by producing q_0 at the price p_w.

Now consider the impact of an IAS on the industry. Once again, necessary changes to the production process shifts the AVC and MC curves of a typical producer up to AVC* and MC*, respectively. Growers still receive the world price p_w, but it is now only economic to produce q_1, at which they accrue the producer surplus $IJKp_w$. Therefore, if the probability of the IAS entering the growing region and becoming established is z, the expected loss of producer surplus at the farm level is $z \times (REFp_w - IJKp_w)$.

The aggregate effect of the IAS across the industry is a contraction of the supply curve in the right-hand frame of Figure 3.4 to S*. In a closed market situation this would result in a domestic market price of p_d, but if there are no barriers to trade the industry can continue to export its product and earn a higher price, p_w. The heavily shaded area GHC indicates total producer surplus. Consumer surplus is unaffected since the price remains at p_w, so is still MNC. Hence, expected loss of producer surplus at the industry level can be expressed as $z \times ABHG$.

As previously, any response costs incurred by industry or government (RC) must also be added to the lost producer surplus to estimate the extent of losses. Formally, the expected total damage cost to an exporting industry (C_z^X) associated with an IAS with an entry probability of z can be calculated as:

$$C_z^X = z \left[\left(p_w - \int_0^{q_4} S \right) \cdot dq - \left(p_w - \int_0^{q_2} S^* \right) \cdot dq + RC \right]. \quad (3.7)$$

Equation (3.7) states that given an IAS arrival probability of z, C_z^X is equal to the expected difference between the host industry producer surplus without the IAS and the producer surplus with the IAS plus response costs.

Note that had the contraction in supply induced by the IAS been more severe, it could have spelled the end of all exports from the industry. If, for instance, the with-IAS supply curve resembled S**, all exports would cease. The industry could still supply q_3 to the domestic market, but it would only earn a producer surplus of LMC. Sales of $q_4 - q_3$ would effectively be lost as a result of losing IAS area freedom. Note also that at the farm level, such a dramatic cost increase may be sufficient to push individual growers out of the market if the minimum value of their AVC function exceeds p_w.

3.2.4 *Monopolistic competition*

In this section, we again describe the economic impact produced by IAS on a host plant product that is exported, but this time in a market containing where growers have some influence on price. There remain a large number of producers and consumers, but the goods produced are not homogenous and are consequently imperfect substitutes for one another. They may be marketed as a brand emphasising particular features that distinguishes products from those of competitors. These might include physical features of the product or its packaging, the methods used to manufacture or distribute the good, or associating it with lifestyle or consumer trends through advertising. This type of market structure is known as monopolistic competition.

3.2.4.1 *An exporting host industry*

Consider firstly the situation where growers within a region can supply a product relatively cheaply and sell it to other countries at a higher price than they can achieve domestically. We will use the partial equilibrium framework to examine the impacts of an IAS, but this time we will focus only on the aggregate market assessment rather than the individual grower. Assume the domestic market for the good concerned is characterised by a downward sloping demand curve, D, and an upward sloping domestic supply curve, S. The price corresponding to the point at which the demand and supply curves intersect, p_1, represents the equilibrium domestic market price. This situation is depicted in Figure 3.5.

When a host product can be transported, screened for IAS and made available (or "landed") for sale in another country or region at a price higher than p_1 but a lower price than foreign producers are capable of supplying it, there are gains from trade. Growers gain from selling their product at a price higher than they could achieve on the domestic market and foreign consumers gain from having product landed in their country at a cheaper price than local producers can achieve.

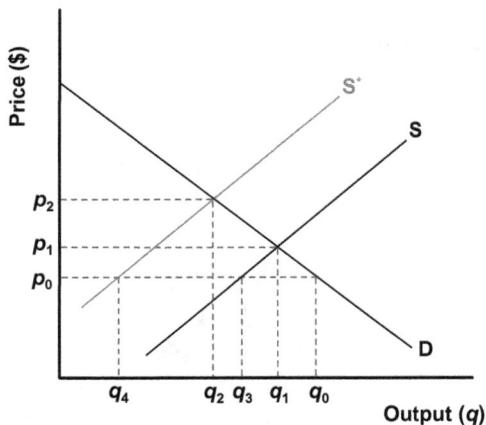

Figure 3.5. The economic impact of an IAS — monopolistically competitive market

If an IAS arrives in the region producing the good, growers would have to carry out on-farm IAS management activities, and therefore must use additional inputs to achieve each unit of output. The domestic supply curve would consequently shift to the left as a lower quantity of host plant products would be offered for sale at a given price. Assume the IAS arrival probability is z and the response costs following its detection are RC, as in the previous section.

Recall that producer surplus is represented by the area below prevailing market price line (i.e. p_2 with and p_1 without the IAS) and above the supply curve (i.e. S* with and S without the IAS). The expected total damage cost to domestic producers within the host plant industry (C_z^X) associated with an IAS with an entry probability of z is therefore:

$$C_z^X = z \left[\left(p_1 - \int_0^{q_1} S \right) \cdot dq - \left(p_2 - \int_0^{q_2} S^* \right) \cdot dq + RC \right]. \quad (3.8)$$

Equation (3.8) states that given an IAS arrival probability of z, C_z^X is equal to the expected difference between the host industry producer surplus without the IAS and the producer surplus with the IAS plus response costs.

Since the IAS raises production costs and lowers supply, the domestic market price rises relative to the export price. If p_2 exceeds the export price, exports will cease altogether. If cost increases more subtlety such that p_2 remains below the export price, export volumes will remain positive. However, given that area freedom from the IAS has now been lost additional phytosanitary measures could be imposed by IAS-free trading partners, raising the landed price of the host product.

3.2.4.2 *An importing host industry*

When the domestic price of a good is above the world price the situation is a little different. Trade may not necessarily be free-flowing. For instance, if goods supplied to the market from non-domestic sources are required to undergo phytosanitary treatments to prevent the transfer of IAS, the landed price will rise according to treatment costs. In some cases these treatment costs may be high enough to prevent non-domestic producers from supplying sufficient quantities to significantly affect the prevailing market price. However, if the pre-import measures are less costly, non-domestic suppliers can exert competitive pressure on domestic suppliers, placing downward pressure on the price.

To demonstrate this diagrammatically, let us assume imported host products are landed in the market depicted in Figure 3.5 at price p_0. Foreign suppliers would supply the quantity $q_0 - q_3$ while domestic growers supply q_3. Given they still face the risk of an IAS arriving, import-competing growers face both a lower price and the risk of increased production costs relative to their non-domestic competitors. If the IAS were to enter and the supply curve shifts to $h(q)$ domestic growers would only be in a position to supply the market with q_3. The shortfall in demand $(q_0 - q_4)$ would be met by foreign suppliers. Using the framework of Figure 3.5, the expected total damage cost to import-competing domestic producers (C_z^I) with an IAS arrival probability z is:

$$C_z^I = z \left[\left(p_0 - \int_0^{q_3} S \right) \cdot dq - \left(p_0 - \int_0^{q_4} S^* \right) \cdot dq + \text{RC} \right]. \quad (3.9)$$

Equation (3.2) states that given a probability of IAS arrival z, C_z^I is equal to the expected difference between the producer surplus if the volume of imports is positive and no outbreak occurs and the producer surplus if an outbreak occurs plus the expected cost of incursion response

3.3 A Dynamic Model

3.3.1 *Relating supply curve shifts to uncertain biological processes*

There is tremendous uncertainty about the processes and circumstances involved in IAS incursion scenarios, as we saw in Chapter 2. In some cases, a species may arrive and be detected early, while in others it may not be detected for several time periods. Sometimes multiple incursion events will occur within a relatively short space of time, while in other cases incursions may be isolated events. Similarly, the manner in which affected growers and collective industries respond and adapt to IAS can be hard to predict.

It follows that the stylised model of Figure 3.1 in which a concise movement of the supply curve is observed does not really do justice to the uncertainty inherent in the system. While it can be used to present specific producer surplus losses, the likelihood of these scenarios actually occurring will seldom, if ever, be known. Moreover, experimentation cannot reduce this uncertainty to the point where scenarios become definite.

So, where does this leave us? Well, rather than use a model with a *certain* supply curve shift induced by an IAS, a decision support analyst might instead adopt a multi-disciplinary approach and consider *uncertain* anthropocentric and ecocentric drivers of change. To do this requires a characterisation of the biology and farming systems involved using sufficient detail to capture a broad range of possible IAS incursion scenarios. Adding to the complexity, when IAS are detected in a region and a response is initiated by government and industry, the cost and outcome of responses is not consistent.

In the remainder of this chapter we will construct a general method of calculating IAS costs that take on-board these complications and condense relevant information for policymakers. It may seem complicated in places because IAS can have so many different impacts, and if we are to account for them in a generic model it needs a certain amount of flexibility. Keep in mind though that the ultimate goal of the model is quite simple. It needs to estimate costs over time so that different IAS can be compared and contrasted. If it can do that, it will at least have formed an important piece of the jigsaw puzzle facing biosecurity policymakers choosing how to best spend their scarce resources.

3.3.2 *A suggested approach*

To move from the aspatial, static framework of Figure 3.1 to a spatiotemporal framework capable of estimating C_z over time for several case studies, we can use an ecological model like the reaction-diffusion model developed in Chapter 2 to simulate the arrival and spread of an IAS in host crops over time. As in Section 2.1, host crop growing areas are also denoted i in this section and the costs associated with an IAS in within that crop are termed d_{it}. Costs d_{it} are predicted as a function of cost and revenue changes that result from an IAS becoming established and spreading.

Formally, we can define d_{it} as the present value of predicted production costs induced by an IAS in host i in time t.[1] The variables taken into consideration when calculating depend on the extent of the incursion and if an eradication response is in place. We will deal with the case of eradication specifically in Chapter 6 where we consider if and when this constitutes the best response. For now though, given the response plans in place within Australia's biosecurity apparatus at the moment, we assume eradication will be attempted when the IAS is first discovered but abandoned if spread continues beyond a specified area.

The first point to keep in mind is that the earlier detection takes place and an eradication attempt is made, the greater the chances of successfully removing the IAS. It follows that the value of d_{it} is influenced by the cost of actually removing the IAS from each host crop and the probability of eradication success. This probability of eradication decreases dramatically with the length of time between IAS arrival and detection.

In constructing our generic model, we assumed the probability of eradication success decreases exponentially with the area affected. That is, the chances of success are high immediately following arrival but fall very quickly thereafter if the IAS remains. We can simulate this using a binomial distribution which returns a zero (i.e. eradication fails) or a one (i.e. eradication succeeds). The chances of the distribution returning a one decreases at an average rate of $e^{-\varphi_i A_{ti}}$, where φ_i is the exponential rate of decline of eradication success probability with respect to area affected in host i, and A_{it} is the area affected by an IAS in host i in time period t weighted by the probability of IAS arrival, establishment and the density of infestation/infection. We will use the

---

[1] The present value of costs or benefits is their discounted value. We will discuss this in Chapter 4.

reaction–diffusion model of Chapter 2 to provide estimates of A_{it} in the case studies to follow, but there are many different spread functions that can be used in practice.

Box 9. Response Plans and Cost Sharing in Australia

When an IAS breaches border defences in Australia, pre-arranged response plans dictate actions that must be taken if the species concerned threatens agricultural industries. The Australian Emergency Plant Pest Response Plan (known as PLANTPLAN) provides an organised and integrated plan to be adopted by all those organisations called upon to respond to an IAS outbreak of significance to plant health, including national, State and local jurisdictions. If eradication is both technically and economically feasible, Australia has a cost sharing arrangement in place to fund the necessary activities. The *Emergency Plant Pest Response Deed* places IAS in one of four cost sharing categories relating to their potential impact on public resources and private industries (Plant Health Australia, 2005). The category dictates an appropriate funding contribution of public and private parties to the eradication. Since it is only eradication responses that are covered under this agreement, there is an incentive to overstate the technical and economic case in favour of eradication to spread the burden of costs. However, if it is deemed that the IAS has spread too far or is too difficult to remove the responsibility for further management largely shifts to State government departments of agriculture, local government and private industries affected by the IAS. Containment and control activities taking place on public land can be provided or subsidised by State governments, but tend to favour IAS with agricultural impacts while environmentally significant IAS continue to spread (Cook *et al.*, 2011b). On private land, control activities tend to be underprovided since the benefits flow on to parties that have not contributed (i.e. positive externality), creating a disincentive for control providers.

By including an early 'knee jerk' eradication response, our model must calculate costs (i.e. d_{it}) differently depending on its success or failure. At any given point in time, t, the costs of an IAS in a host industry, i, can be expressed as the piecewise function:

$$
d_{it} = \begin{cases}
E_{it}A_{it} + \beta_{it} \sum_{i=1}^{n} X_{it} & \text{if } A_{it} \leq A_{it}^{\text{erad}} \\
Y_{it}P_{it}A_{it} + V_{it}A_{it} + \beta_{it} \sum_{i=1}^{n} X_{it} & \text{if } A_{it} > A_{it}^{\text{erad}}.
\end{cases} \tag{3.10}
$$

Here, E_{it} is the cost of removing an IAS from a given area in host crop i in year t; A_{it}, is the area affected by an IAS in host i in year t weighted by the probability of arrival and density of infestation/infection (recalling Equations (2.1)–(2.4)); β_{it} is the percentage reduction in export earnings attributable to a loss of IAS area freedom in host i in year t; $\sum_{i=1}^{n} X_{it}$ is the present value of exported host products of host i in year t; A_{it}^{erad} is the maximum technically feasible area of eradication in host i in year t; Y_{it} is the mean change in yield resulting from an IAS becoming established in host crop i in year t; P_{it} is the prevailing price for an affected host in year $t - 1$; and V_{it} is the increase in variable cost of production per unit of area induced by IAS management methods in host i in year t.

Table 3.1 lists the parameters needed to estimate d_{it} for each of the case studies that follow in Chapter 4. The remainder of this section describes how each of these parameters can be used to simulate arrival, dispersal and response effort within our model.

The total damage inflicted by an IAS is a function of the area affected, host price, yield reduction, control cost, gross value of production and the value of exports affected. Recall from Chapter 2 that our prediction of A_{it} is inclusive of an organism's entry and establishment probability in region i, z_i, and therefore represents the area predicted to be in need of additional management effort due to its presence (Equation (2.4)).

Linking changes in aggregate output of a host industry to IAS abundance can be a challenge. This is not as much of a problem in perfectly competitive market structures (i.e. where price remains constant), but is important given that the amount produced affects the price of a host commodity in monopolistically competitive markets. For practical purposes, we use predicted production loss

Table 3.1. Parameters for a model to assess the spread and impacts of invasive alien species over time

Description and equation symbol	Units
Area currently affected, A^{min}	ha
Average cost of eradication, E	$ per ha
Demand elasticity, ε	unitless
Exponential rate of decline for eradication success probability with respect to area affected, φ	%
Gross value of production of A^{max} divided by 100, G	$
Increased variable cost of production if eradication fails, V	$ per ha
Intrinsic rate of population/infection density increase, δ	year^{-1}
Intrinsic rate of affected area increase, r	year^{-1}
Intrinsic rate of satellite generation per unit area of infection/infestation, μ	#/ha
Maximum area affected, A^{max}	ha
Maximum area considered for eradication, A^{erad}	ha
Maximum population/infection density, K	# per ha
Maximum number of satellite sites generated in a single time step, s^{max}	#
Minimum population/infestation density, N^{min}	# per ha
Minimum number of satellite sites generated in a single time step, s^{min}	#
Population/infection diffusion coefficient, D	ha per year
Prevailing price of affected commodity in the first time step, P_0	$/unit
Probability of entry and establishment, z	%
Reduction in export earnings attributable to a loss of pest/disease area freedom, β	%
Yield reduction despite control, Y	%

$Y_{it}A_{it}$ as a proxy for the reduction in output when dealing with IAS impacts in these type of markets. In reality, the yield loss is only part of the story. To truly gauge a producer's output response and their resilience to shocks like IAS outbreaks requires detailed production cost information. However, since agricultural markets are often highly competitive, each producer has an incentive withhold their technical and cost information to prevent others from mimicking them and eroding their share of the market. As a consequence, MC information is rarely made available.

Box 10. Spread Model Logic

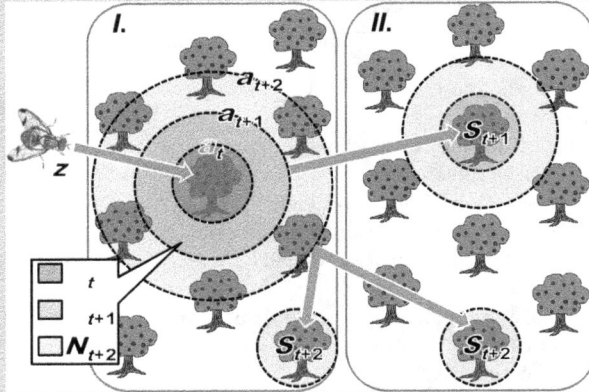

It may be easier to explain how the spread model works with the aid of an example. In the diagram above there are two representative fruit production areas, labelled I and II, threatened by a fruit fly species. Female fruit flies inject eggs below the skin of host fruit where they hatch and the larvae feed on the pulp, rendering it useless for commercial purposes. Hence, many fruit fly species present biosecurity risks. The probability that area I will experience an incursion is z. When an incursion does take place, the time period in which it occurs is labelled t and the diagram shows the extent of spread during this and two subsequent time periods (i.e. t, $t+1$ and $t+2$). At the initial infestation site, spread is assumed to occur at the same average rate in every direction. The area affected is calculated by the model at the end of each time step, so increases as a series of concentric circles labelled a_t, a_{t+1} and a_{t+2}. The density of infestation within each circle also increases over time, shown as N_t, N_{t+1} and N_{t+2}, respectively. In time step $t+1$ the infestation jumps from production area I to II where it begins to spread. This new 'satellite' site is labelled S_{t+1}. By $t+2$, two more satellite sites appear, both labelled S_{t+2} (Cook and Fraser, 2014).

Hence, $Y_{it}A_{it}$ is used as a simple output estimate and combined with the lagged host price (i.e. host price in the previous time step), P_{it-1}, to calculate $P_{it} = P_{it-1}[1-(\frac{Y_{it}A_{it}}{G_{it}\varepsilon})]$. Here, G_{it} is the gross value of production divided by 100 and ε is the price elasticity of demand for the affected host i. Note that where the market in question is perfectly competitive and producers face a perfectly elastic demand curve (i.e. $\varepsilon = \infty$), price remains constant over time (i.e. $P_{it} = P_{it-1}$).

Given P_{it}, we can calculate the reduction in producer surplus associated with the new lower level of output

---

Box 11. Price Elasticity of Demand

The price elasticity of demand, ε, indicates how the quantity demanded of a product changes as its price changes. It is calculated as the ratio of percentage change in quantity demanded over the percentage change in the price. That means that if ε is equal to zero the amount of the good demanded by consumers does not change at all when its price changes. This rare type of good is said to have *perfect price inelasticity*. If, on the other hand, ε has a value between 0 and 1 demand for the good changes less than proportionally with its price. This type of good is termed *price inelastic*. If ε equals one, this tells us that the demand for the good will change exactly proportionally to a change in price. This good is said to have *unit price elasticity*. If ε is greater than one, demand is affected more than proportionally by a change in price, and the good is said to have an *elastic* demand. Finally, if ε is infinite (∞), the good has a perfectly elastic demand curve and any change in price will result in demand falling to zero.

Values of the cost/loss parameters E_{it} (i.e. the cost of eradication), Y_{it} (i.e. the value of yield reduction), V_{it} (i.e. variable production costs) and β_{it} (i.e. percentage reduction in export revenue caused by a loss of area freedom) as they relate to specific IAS are detailed in the reports for each case study in Chapter 4.

By summing costs across all hosts n in a region we can estimate the total damage cost, C, induced by the IAS:

$$C = \sum_{i=1}^{n} d_{it}. \tag{3.11}$$

This indicator C is what we will use to communicate to policymakers the likely market-based impact of an IAS. We will look at the dynamics of this impact for different IAS in Chapters 4 and 5.

Before we do, however, it is important that we remain grounded and recognise that, as with any model output, C is a simplification of reality and is of the same overall quality as the information used to calculate it. This is especially important to keep in mind when dealing with complex agro-ecological systems that are inherently uncertain. Of course, it is also important to consider that the value of models does not just rest with their outputs. They can be important mind maps that can help decision makers and analysts alike to think about complex problems and how to adapt to them.

3.4 Uncertainty and Model Parameterisation

From the beginning of this book we have acknowledged that we will rarely be in the position of having access to all the information we want to be able to make accurate predictions about the future. Indeed, if we did there would not be an IAS externality problem in the first place. The reality is though that empirical evidence about our model parameters and the impacts of management decisions is less than perfect and decision-support analysts rarely have the time or resources to solicit the information required. Moreover, when data is available there are issues concerning its collection and consistency with other data sources. This can often mean that the analyst is 'flying blind', forcing them to look beyond the available data.

As stated in Chapter 2, when empirical evidence is lacking it is common practice to use direct or indirect expert elicitation approaches to fill the data gaps (Burgman, 2005; Low Choy *et al.*, 2009; Kuhnert *et al.*, 2010; Martin *et al.*, 2012). Direct elicitation requires experts on a particular species to express their knowledge in terms of quantities required by analysts. For instance, experts may be asked to provide statistical summaries (e.g. a lower bound, a best estimate, and an upper bound) or a full parametric probability distribution. In contrast, indirect elicitation asks experts questions related to their experiences that are then encoded into the quantities needed. For example, the expert may be asked about expected IAS costs given different levels of abundance, which the analyst then translates into a corresponding probability distribution for a model parameter. This indirect approach tends to be more comfortable for the experts, although not necessarily for the decision-support analyst (Allan *et al.*, 2010).

From Equation (3.10), d_{it} includes parameters which cannot be given definitive values. These uncertain parameters are specified within the model as distributions and a Monte Carlo simulation used to sample from each distribution using the $@Risk^{TM}$ software package (Palisade Software, Ithaca, New York). In each of the 10,000 model iterations, one value is sampled from the cumulative distribution function so that sampled parameter values are weighted according to their probability of occurrence. Model calculations use the sampled set of parameters.

Types of distributions used in the model include: (i) PERT, mentioned previously and specified using expert opinions of minimum, most-likely and maximum values; (ii) uniform, a rectangular distribution bounded by minimum and maximum values; (iii) binomial, returning a zero (i.e. failure) or one (i.e. success) based on a number of trials and the probability of a success — recall we used this to represent eradication success or failure in the previous section; and (iv) discrete, a distribution in which several discrete outcomes and their probabilities of occurrence are specified.

When using experts to form PERT distributions, we must acknowledge that the process will be subject to heuristics and biases. Individual experts use both a spatial proximity heuristic (e.g. they attribute relatively more damage to IAS in their home country than elsewhere) and an effect heuristic (e.g. tangible economic impacts are regarded as more severe than equivalent environmental impacts) (Dahlstrom Davidson *et al.*, 2013). In a group elicitation setting, there are added concerns introduced via dominant group members. To overcome these limitations, structured approaches such as the Delphi method can help in allowing experts to adjust their own estimates in light of the answers of other group members while maintaining anonymity (De França Doria *et al.*, 2009).

Risk analysis has become a discipline in its own right and readers should consult relevant texts for specific information about different estimation techniques. It is increasingly popular to use Bayesian belief networks and probability bounds analysis to develop models of complex problems. Here, we simply highlight the need to communicate uncertainty to policymakers rather than a critique of these methods. Suffice it to say it has never been easier to depart from simple deterministic representations of complex IAS problems.

3.5 Conclusion

In this chapter we have added to our predictive model of area affected by estimating the cost and revenue changes that will probably result over time. With so much uncertainty in the way IAS spread, reproduce and cause change in the system, we must be under no illusion about the difficulty of predicting resultant financial impacts. While we can never say definitively what will happen and when, we can use our model to at least think about the problem in an appropriate perspective. A model is necessarily a simplification of reality and should never be thought of as perfect, and we will discuss ways we can bolster imperfect models in Chapter 6. However, in the next chapter we will use the model on its own to examine several case studies and see what we can deduce about the economic impacts of these IAS on different agricultural industries. The geographic region we will use is WA, but remember our models are generic and could just as easily be applied elsewhere.

CHAPTER 4

CASE STUDIES OF IAS AFFECTING PLANT INDUSTRIES

4.1 Introduction

The model we developed in Chapter 3 captures key economic drivers that enable us to represent IAS market impacts over time, but this is by no means the end of the story. In some ways, it is just the beginning. Condensing the model results into small bites of information for "time-poor" policymakers is a challenge in and of itself. Information about total damage cost (C, in Chapter 3) can be expressed in different ways, and decision-making groups may have their own thoughts and opinions about how this is best presented to make it easily understood.

In this chapter, a set of indicators are suggested that we feel are useful when communicating the scale and dynamics of (potential) incursion events to diverse groups of stakeholders. In the case studies we present, total damage cost information is given for two separate scenarios; one in which an incursion takes place in time period one and the other in which outbreaks occur randomly over time with a frequency dependent on the probability of arrival. For each IAS example, a large number of iterations are used to explore the full extent of uncertainty in C calculated over a 30-year period under both scenarios. Examples of IAS are separated into four basic taxonomic groups: Terrestrial invertebrates (Section 4.2); plant pathogens (Section 4.3); vertebrates (Section 4.4); and terrestrial plants (Section 4.5). By using such a broad set of species, we hope to demonstrate the flexibility of the model and the information it can provide to help policymakers.

In reporting results, box whisker plots are used to summarise model output data in terms of area affected and impact over time, and the spread of that data. Each plot shows the 25th percentile, the median (i.e. the 50th percentile), the 75th percentile, and remaining values up to and including the 5th and 95th percentiles of the distribution of model output. The boundaries of the boxes indicate the 25th percentiles and the 75th percentiles. From the length of the box, the variability can be determined (i.e. the larger the box, the greater the spread of model predictions and the more uncertain the results). The horizontal lines inside the boxes represent the medians. If a median is not in the centre of a box, the distribution is skewed. Values lying between the 5th and 25th percentiles and between the 75th and 95th percentiles are by the lines extending from the top and bottom of each box. These lines are called whiskers. If the upper whisker is longer than the lower whisker, it implies a positive skewness, and vice versa.

When viewing the cost information over multiple time periods it is important to keep in mind that future costs are discounted. Discounting has an erosive effect on costs that increases with time, and as such future costs can be seen to fall in real terms over successive time periods irrespective of IAS prevalence. For simplicity, we use a traditional exponential discounting approach with a constant discount rate of 5% per annum.

Note also that in early time steps there can be a lot of uncertainty about the total damage costs involved if an eradication response is initiated. In rare instances this will be successful, but in most cases it will not lead to the long term removal of the infestation. Moreover, response costs will vary according to a range of factors, including the size of the initial incursion and the lag between introduction and detection. Both of these are variable in the model. In subsequent years when eradication is likely to have ceased there is less uncertainty in predicted costs.

Box 12. What is Discounting?

Assume, for example, a farmer is spared $100 worth of crop damage from IAS in one time period. In the following time period, the farmer has the option of putting this additional revenue back into cropping, or to invest it in something else. She may choose to put the money into stocks, shares or bonds and earn an interest rate of, say 10% per annum. By the end of the initial time period in which the $100 was spared the interest payment that could be earned by putting it in the bank is an opportunity cost of reinvesting in cropping. So, if the same crop damage was prevented in the second time period it would only be worth $90, instead of $100. By the end of the third time period it is only worth $72.90, and so forth. With a private discount rate of 10%, the initial $100 is effectively worthless by the end of the 14th time period.

To further condense model information about C, we annualise the costs over time using a simple average. While there are problems with outlying data points skewing results, in this chapter we simply use an annualised cost indicator as an example.[2] The distributions of annual average C values predicted by the model are shown as relative frequency histograms. The average annual cost histograms for the random incursions scenario (listed as SCENARIO in the figure legends) are shown using clear bars and are overlaid with results from the incursion year 1 scenario (listed as CONTROL CASE), shown as shaded bars. The data are separated into categories or 'bins' along the horizontal axis according to the value of costs calculated by the model. The relative frequency diagrams show the probability (on the vertical axis) of the model producing values in any given bin. The longer the vertical bars associated with a bin, the greater the likelihood of the model outputting a result in that range.

---

[2]While we briefly return to the general issue of communicating uncertain information regarding low probability and high consequence events to policymakers in Chapter 8, this remains a considerable challenge given differences in measured and perceived risks.

4.2 Terrestrial Invertebrates

4.2.1 *Outline*

The State of Western Australia (WA) remains free from many IAS found elsewhere in the world partially as a result of its isolation. However, protecting the region from future incursions remains a challenge given the 2,889 kilometre-long coastline and 1,862 kilometre-long interstate border to police. Potential pest and disease entry pathways continue to increase along with the growing volume of people and produce entering the State. One of the most challenging groups of species to keep out is terrestrial invertebrates. Not only are they small and difficult to detect, they are also highly mobile with a tendency to reproduce quickly. They are versatile animals with multiple life stages, some which involve dormancy for extended periods. Added to this is their ability in some cases to act as vectors for viruses and plant pathogens that add to the direct impacts they themselves inflict (Baskin, 2002). Given the massive variety in potential invertebrate IAS, the case studies below have been selected in order to highlight the potentially large impact on the WA economy such small organisms might have.

4.2.2 *Cabbage seedpod weevil (Ceutorhynchus assimilis)*

4.2.2.1 *Description*

Cabbage Seedpod Weevil (*Ceutorhynchus assimilis*) is a pest of oilseed rape (canola) crops in Europe and North America. It is native to Europe but became established in Canada and the United States in the 1930s where it now causes serious damage to canola crops (Centre for Agricultural Bioscience International, 2014).

Adult cabbage seedpod weevil overwinter in dry soil, leaf litter or underneath scrub adjacent to crops and migrate to brassicaceous seed crops as they start to flower. Here they feed for three to four weeks on buds, flowers, pods and stem tips before mating. Females bore a hole through the pod wall with their rostrum and insert a single egg. Larvae hatch between 6 and 30 days later, depending on prevailing temperatures, and feed within the pods for 14–40 days. During this time, each larva destroys four to six seeds before it is grown (Ulmer and Dosdall, 2006). Larva then bore an exit hole through the pod wall and drop to the soil to pupate (Centre for Agricultural Bioscience International, 2014).

Direct grower losses attributable to cabbage seedpod weevil are difficult to estimate since damage to canola crops can actually promote growth due to the plants' ability to replace damaged pods (although late damage may lead to immature pods at harvest) (Centre for Agricultural Bioscience International, 2014). However, pod damage may act as a catalyst for secondary infection and promote other seed-feeding invertebrates (Ulmer and Dosdall, 2006). If cabbage seedpod weevil were to become established in WA it is expected to infest all areas in the Southwest where canola is grown, necessitating the use of additional chemical treatments to minimise these types of crop injury.

In addition to direct costs, potential canola export revenue loss is considered to be substantial since cabbage seedpod weevil larvae reside within seed cavities. This threatens the significant growth in canola exports from WA that has occurred in the last five years driven by biodiesel production in Europe. From less than 0.2 million tonnes in 2006/2007, WA produced 0.9 million tonnes in 2011/2012 with a total export value of $0.5 billion (Department of Agriculture and Food, 2012; ABS, 2014). In 2013/2014 this is estimated to have risen to just over $1.0 billion, giving the State a total share of over 7% of the global market (Paterson and Wilkinson, 2015). Almost 90% of WA canola is sold to the European Union (EU) where cabbage seedpod weevil is endemic, but could be subject to price discounting if area freedom is lost (Department of Agriculture and Food, 2012). Exports to other markets that are free of the pest, such as South Korea, Japan and Malaysia, would be more severely affected. In the absence of definitive information, we estimate the export losses of between 20% and 30% of their current level (i.e. uniform $(-0.1, -0.3)$).

4.2.2.2 *Parameters*

See Table 4.1.

Table 4.1. Cabbage seedpod weevil model parameters

Description	Nil management
Area currently affected, A^{min} (ha).	0
Cost of eradication, E ($/ha)	Uniform(5.0×10^5, 1.0×10^6)
Demand elasticity, η.[a]	∞
Exponential rate of decline for eradication success probability with respect to area affected, φ	Pert(-0.2, -0.15, -0.1)
Increased variable cost of production if eradication fails, V ($/ha).[b]	0
Intrinsic rate of population/infection density increase, δ (yr^{-1}).[c]	Pert(0.20,0.35,0.50)
Intrinsic rate of infestation/infection area growth, r (yr^{-1}).[c]	Pert(0.20,0.35,0.50)
Intrinsic rate of satellite generation per unit of area affected, μ (#/ha).[c]	Pert(1.0×10^{-3}, 5.5×10^{-3}, 1.0×10^{-2})
Maximum area affected, A^{max} (ha).[d]	1.1×10^6
Maximum area considered for eradication, A^{erad} (ha)	Pert(5.0×10^3, 7.5×10^3, 1.5×10^3)
Maximum infestation/infection density, K (#/ha).[c]	Uniform(1.0×10^3, 1.0×10^4)
Maximum number of satellite sites generated in a single time step, s^{max}(#).[c]	Pert(10,15,20)
Minimum infestation/infection density, N^{min} (#/ha).[c]	Pert(0.00,0.025,0.050)
Minimum number of satellite sites generated in a single time step, s^{min} (#).	1.0
Population/infection diffusion coefficient, D (ha/yr).[c]	Pert(2,3,4)

(Continued)

Table 4.1.   (*Continued*)

Description	Nil management
Prevailing price for affected commodities in the first time step, P_0 ($/T).[d]	Canola 542
Probability of detection (%).	Pert(20,40,60)
Probability of entry and establishment, z (%).[e]	Uniform(7.0×10^{-2}, 5.0)
Reduction in export earnings attributable to a loss of pest/disease area freedom, β.[e]	Uniform(20,30)
Yield reduction despite control, Y(%).[f]	Uniform(15,35)

[a]Host products are predominantly sold to export market and WA is not sufficiently large to exert pressure on the world price; [b]Two applications of alpha-cypermethrin (or equivalent pyrethroid) at 0.2–0.3L/ha or $1.40–2.10/ha (Rural Solutions SA, 2014); [c]Waage *et al.* (2005); [d]ABS (2012); [e]Cook (2003); [f]Homan and McCaffrey (1993) and Brodeur *et al.* (2001).

4.2.2.3 *Results*

See Figures 4.1–4.3.

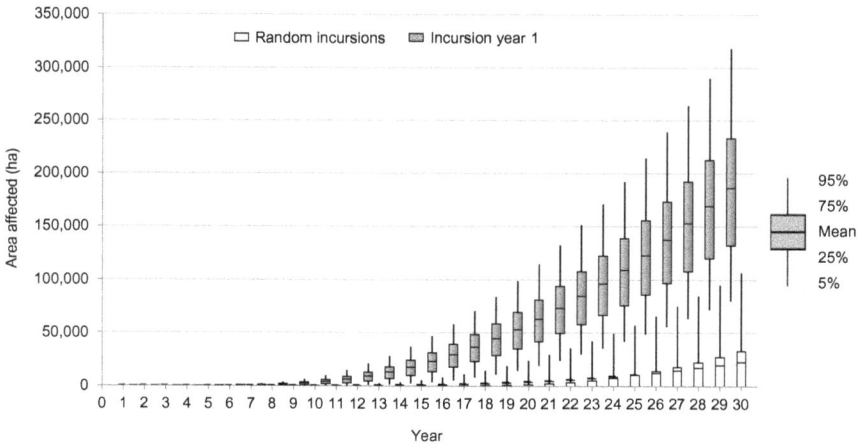

Figure 4.1.   Area affected by cabbage seedpod weevil in WA if an incursion occurs in year 1 and if incursions occur randomly over 30 years

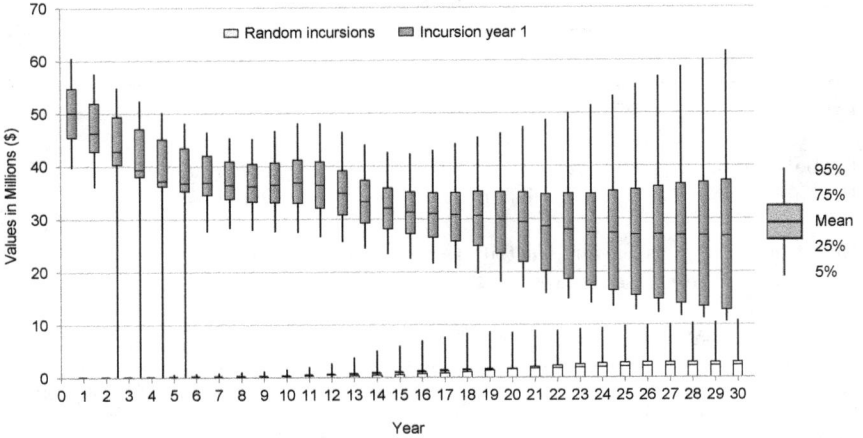

Figure 4.2.   Total damage cost of cabbage seedpod weevil in WA if an incursion occurs in year 1 and if incursions occur randomly over 30 years

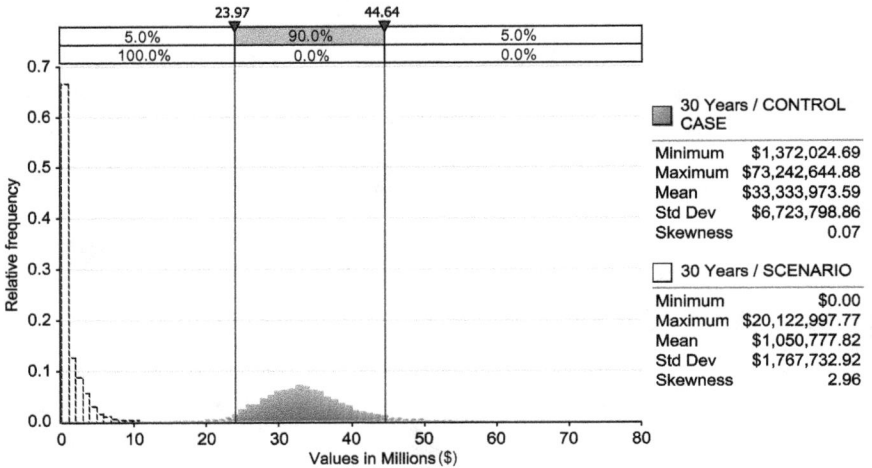

Figure 4.3.   Annualised total damage cost of cabbage seedpod weevil in WA if an incursion occurs in year 1 and if incursions occur randomly over 30 years

4.2.2.4 *Conclusion*

Both yield losses and export market losses would result from *C. assimilis* if it were to enter WA and become established. The insect threatens the significant growth in canola exports that has occurred over the last five years. If it were to occur in the State in the current time period it is estimated that total damage costs would average $33.3 million over the subsequent 30 years. However, since the insect has a relatively low likelihood of arriving in WA, random incursions over this period are expected to cause total damage costs of just over $1.0 million per year.

4.2.3 *Khapra beetle (Trogoderma granarium)*

4.2.3.1 *Description*

Khapra beetle (*Trogoderma granarium*) is a destructive pest of stored grain and cereal products in hot and dry climates. It is capable of rapid reproduction and the impact of larvae feeding on grain is a serious economic problem wherever the beetle is found. Although it has never been observed to fly, khapra beetle has successfully spread from its native India and is now found throughout Southeast Asia, much of Africa, the Middle East and parts of Europe and South America as a contaminant of traded goods and containers where it may be transported while in diapause (Centre for Agricultural Bioscience International, 2014).

Khapra beetle larvae look very similar to those of other relatively unimportant *Trogoderma* species and are best identified by microscopic examination. Some fumigants give control at high dosages, but the beetle is more resistant to these than most stored product pests. High concentrations of fumigant must be maintained over the fumigation period to allow penetration into all cracks and crevices, making the disinfestation procedure costly.

In 2007, khapra beetle was detected in WA at a home in the Perth suburb of Scarborough. The beetle was found in packing boxes and personal belongings shipped from overseas. The *Emergency Plant Pest Response Deed* (Plant Health Australia, 2005) took effect and the home was immediately fumigated. Subsequent detections in WA have not occurred. A national trapping programme carried out in collaboration with Cooperative Bulk Handling, GrainCorp and Viterra Inc. at 91 sites across Australia between October 2009 and October 2011 failed to detect the insect (Cunningham, 2012).

If it were to again be discovered in WA and eradication was unsuccessful, khapra beetle is expected to survive comfortably throughout grain growing areas of the State and probably spread to eastern States over time (Banks, 1977). Direct losses to grain in affected storage facilities could be 5–20% (Ahmedani *et al.*, 2011). It is also likely that some trading partners would impose phytosanitary measures on grain imported from WA, and possible Australia to the extent that it is considered one trading block by some importing countries. Resultant export losses would probably be in the order of 10–20% (McElwee, 2000).

4.2.3.2 *Parameters*

See Table 4.2.

Table 4.2. Khapra beetle model parameters

Description	Nil management
Area currently affected, $A^{\min}$ (ha).	0
Cost of eradication, E ($/ha).[a]	Uniform(1.0×10^6, 1.0×10^7)
Demand elasticity, η.[b]	∞
Exponential rate of decline for eradication success probability with respect to area affected, ϕ.	Pert(-0.2, -0.15, -0.1)
Increased variable cost of production if eradication fails, V ($/ha).[c]	0
Intrinsic rate of population/infection density increase, δ (yr^{-1}).[d]	Pert(0.20, 0.35, 0.50)
Intrinsic rate of infestation/infection area growth, r (yr^{-1}).[d]	Pert(0.20, 0.35, 0.50)
Intrinsic rate of satellite generation per unit of area affected, μ (#/ha).[d]	Pert(5.0×10^{-2}, 7.5×10^{-2}, 1.0×10^{-1})
Maximum area affected, $A^{\max}$ (ha).[e]	6.0×10^6
Maximum area considered for eradication, A^{erad} (ha).	Pert(5.0×10^3, 7.5×10^3, 1.5×10^3)
Maximum infestation/infection density, K (#/ha).[d]	Uniform(1.0×10^3, 1.0×10^4)

(*Continued*)

Table 4.2. (*Continued*)

Description	Nil management
Maximum number of satellite sites generated in a single time step, $s^{max}(\#)$.[d]	Pert(30, 40, 50)
Minimum infestation/infection density, $N^{min}(\#/ha)$.[d]	Pert(0.00, 0.025, 0.050)
Minimum number of satellite sites generated in a single time step, $s^{min}(\#)$.	1.0
Population/infection diffusion coefficient, D (ha/yr).[d]	Pert(1.0, 1.5, 2.0)
Prevailing price for affected commodities in the first time step, P_0 ($/T).[e]	Barley 213 Wheat 280 Oats 176
Probability of detection (%).	Pert(20, 40, 60)
Probability of entry and establishment, z (%).[f]	Uniform(7.0×10^{-2}, 5.0)
Reduction in export earnings attributable to a loss of pest/disease area freedom, β.[f]	Uniform(10, 20)
Yield reduction despite control, $Y(\%)$.[g]	Uniform(5, 20)

[a]Specified with reference to Pasek (1998); [b]Host products are predominantly sold to export market and WA is not sufficiently large to exert pressure on the world price; [c]McElwee (2000); [d]Specified with reference to Waage *et al.* (2005); [e]ABS (2014); [f]Cook (2003) and McElwee (2000); [g]Ahmedani *et al.* (2011).

4.2.3.3 *Results*

See Figures 4.4 and 4.5.

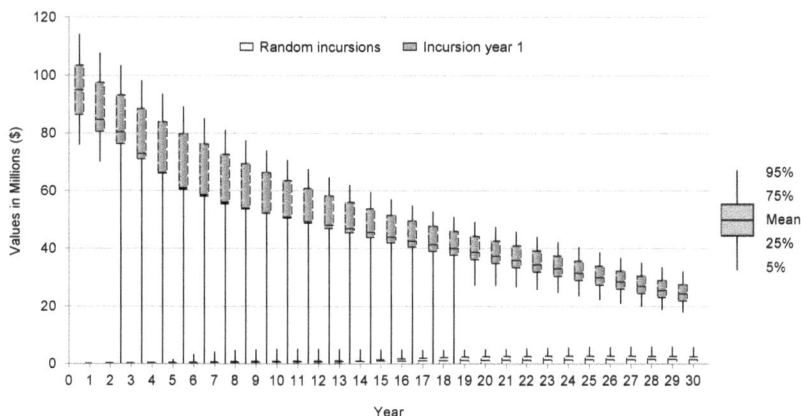

Figure 4.4. Total damage cost of khapra beetle in WA if an incursion occurs in year 1 and if incursions occur randomly over 30 years

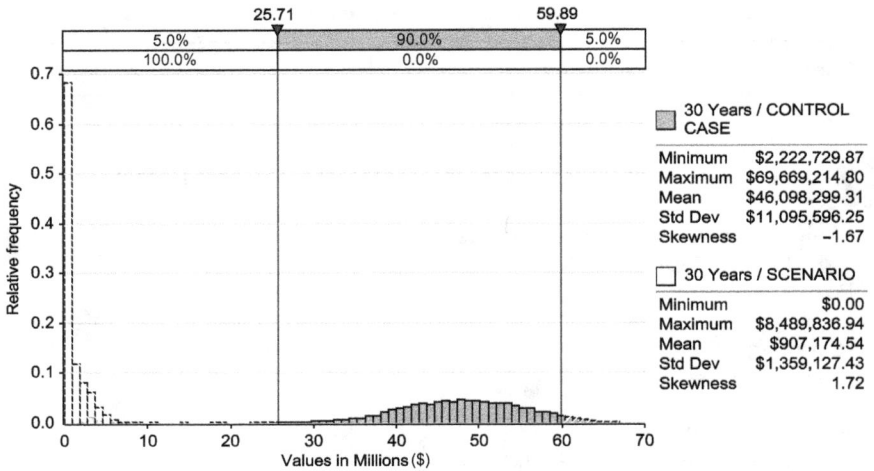

Figure 4.5. Annualised total damage cost of khapra beetle in WA if an incursion occurs in year 1 and if incursions occur randomly over 30 years

4.2.3.4 *Conclusion*

Khapra beetle is a potentially devastating pest of stored grain products currently not established in WA. The results of this analysis indicate that if an incursion were to take place in the current time period, annual losses over the subsequent 30-year period could be as high as $69.7 million, but are more likely to be around $46.1 million. These high costs are mainly attributable to a loss of area freedom disrupting sales of grain to trading partners. In contrast, if we factor in a relatively low probability of khapra beetle entering and becoming established in WA, expected annual losses are considerably lower at approximately $0.9 million per year.

4.2.4 *Wheat stem sawfly (Cephus cinctus* Norton)

4.2.4.1 *Description*

The wheat stem sawfly (*Cephus cinctus*) is an important pest of cereals throughout Europe, Asia, North America and Africa (Weiss and Morrill, 1992). They are not flies at all, but are primitive fly-like wasps. The name 'sawfly' originates from the saw-like ovipositor of adult females (Malipatil and Plant Health Australia, 2008). Yield losses from sawfly larvae feeding on stems of cereal plants can be severe (Özberk *et al.*, 2005, Botha *et al.*, 2004). If wheat stem sawfly were to become established in WA, disruptions to grain exports could also occur if some countries require specific declarations before accepting grain from the State (or Australia as one trading block). However, to date no grain importing countries have issued specific statements regarding area freedom certification from sawflies (i.e. see http://www.daff.gov.au/micor/Plants/Pages/plants.aspx).

Wheat stem sawfly larvae tunnel into the stem of host plants leaving a trail of frass. Upon reaching maturity larvae compact a mound of this frass within their burrows 4–5 cm above the ground. Below this is cut a V-shaped groove around the interior of the culm whilst a thin layer of epidermis remains uncut. Below the groove another plug of frass is placed, the upper surface of which is concave whilst the lower surface is flat. To this lower surface is attached a spun pupal case in which overwintering occurs. The purpose of this elaborate creation is to cause the stem to snap off at the point of the V-shaped incision when blown by wind; thus leaving the pupae safely inside the overwintering chamber. Adults subsequently emerge, forcing their way out through the plug (Centre for Agricultural Bioscience International, 2014; Ries, 1926).

If introduced to WA, wheat stem sawfly is expected to affect approximately 50% of farms cropping wheat (Cook, 2003). Because larvae exist within the stems of infested crops they are difficult to treat with insecticides. Cultural control, biological control and host plant resistance have also proven ineffective against wheat stem sawfly (Weiss and Morrill, 1992). Therefore, the use of chemical against adults is considered most likely mature insects can be eliminated during flights by application of insecticides registered for use in wheat (Centre for Agricultural Bioscience International, 2014), but yield losses of up to 25% may persist over time (Botha *et al.*, 2004).

4.2.4.2 *Parameters*

See Table 4.3.

Table 4.3. Wheat stem sawfly model parameters

Description	Nil management
Area currently affected, $A^{\min}$ (ha).	0
Cost of eradication, E (\$/ha).[a]	Uniform(1.0×10^6, 1.0×10^7)
Demand elasticity, η.[b]	∞
Exponential rate of decline for eradication success probability with respect to area affected, ϕ	Pert(-0.2, -0.15, -0.1)
Increased variable cost of production if eradication fails, V (\$/ha).[c]	Uniform(3, 5)
Intrinsic rate of population/infection density increase, δ (yr^{-1}).[d]	Pert(0.20, 0.35, 0.50)
Intrinsic rate of infestation/infection area growth, r (yr^{-1}).[d]	Pert(0.20, 0.35, 0.50)
Intrinsic rate of satellite generation per unit of area affected, μ (#/ha).[d]	Pert(5.0×10^{-2}, 7.5×10^{-2}, 1.0×10^{-1})
Maximum area affected, $A^{\max}$ (ha).[e]	3.5×10^6
Maximum area considered for eradication, A^{erad} (ha).	Pert(5.0×10^3, 7.5×10^3, 1.5×10^3)
Maximum infestation/infection density, K (#/ha).[d]	Uniform(1.0×10^4, 1.0×10^5)
Maximum number of satellite sites generated in a single time step, $s^{\max}$ (#).[d]	Pert(30, 40, 50)
Minimum infestation/infection density, $N^{\min}$ (#/ha).[d]	Pert(0.00, 0.025, 0.050)
Minimum number of satellite sites generated in a single time step, $s^{\min}$ (#).	1.0
Population/infection diffusion coefficient, D (ha/yr).[d]	Pert(2, 3, 4)
Prevailing price for affected commodities in the first time step, P_0 (\$/T).[e]	Barley 213 Oats 176 Wheat 280

(*Continued*)

Table 4.3. (*Continued*)

Description	Nil management
Probability of detection (%).	Pert(20, 40, 60)
Probability of entry and establishment, z (%).[f]	Uniform(1.5, 21.0)
Reduction in export earnings attributable to a loss of pest/disease area freedom, β.[f]	Uniform(5, 10)
Yield reduction despite control, Y(%).[g]	Uniform(5, 25)

[a]Specified with reference to Pimentel (2014); [b]Host products are predominantly sold to export market and WA is not sufficiently large to exert pressure on the world price; [c]Chlorpyrifos ($10.00/L) applied at 300–500 g/L (Rural Solutions SA, 2014); [d]Specified with reference to Waage *et al.* (2005); [e]ABS (2014); [f]Cook (2003); [g]Botha *et al.* (2004) and Özberk *et al.* (2005).

4.2.4.3 *Results*

See Figure 4.6–4.8.

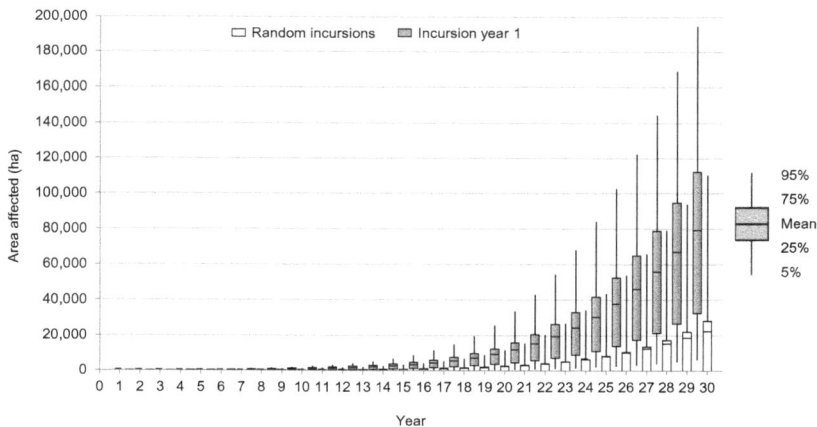

Figure 4.6. Area affected by wheat stem sawfly in WA if an incursion occurs in year 1 and if incursions occur randomly over 30 years

4.2.4.4 *Conclusion*

Despite not having a significant impact on grain exports, wheat stem sawfly has the potential to impose high costs on WA grains producers. The insect is difficult to control and yield losses are expected to

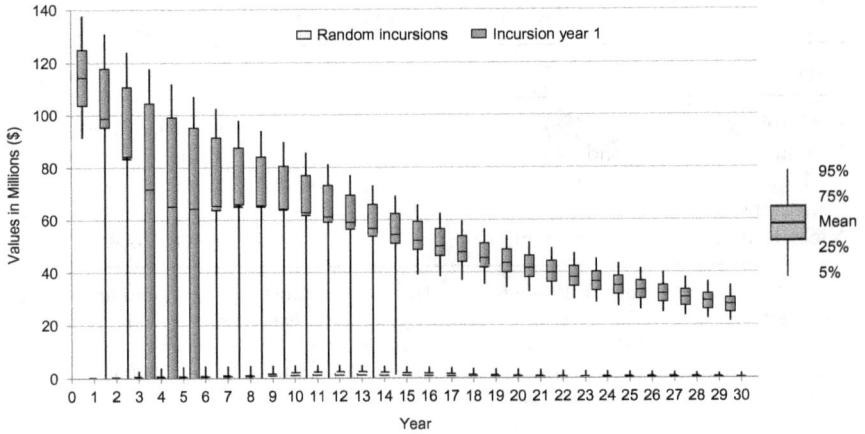

Figure 4.7. Total damage cost of wheat stem sawfly in WA if an incursion occurs in year 1 and if incursions occur randomly over 30 years

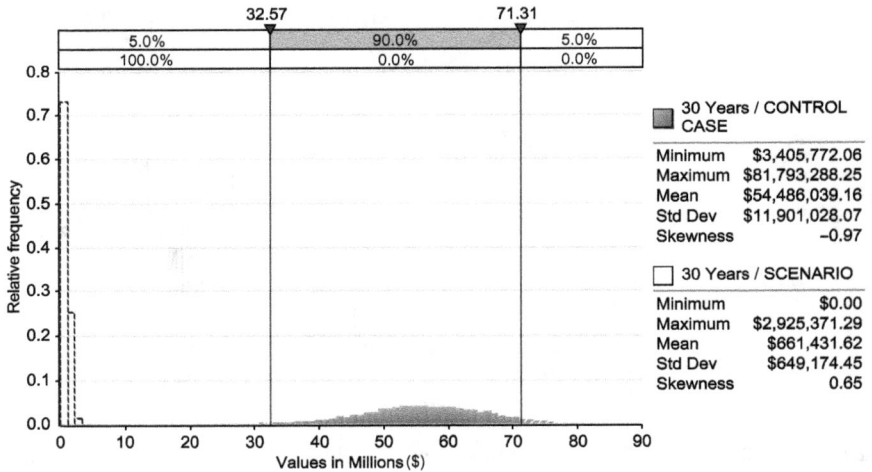

Figure 4.8. Annualised total damage cost of wheat stem sawfly in WA if an incursion occurs in year 1 and if incursions occur randomly over 30 years

persist despite applications of insecticide if the insect were to become established. If establishment took place in year 1, these costs are predicted to average $54.5 million per year over the subsequent 30-year period. If the probability of entry and establishment is taken into account, however, expected total damage costs are around $0.6 million per year.

4.2.5 *Summary — terrestrial invertebrates*

Using the model developed in Chapter 3 with two alternative scenarios (i.e. one when an organism arrives in the State in the current year, and another in which arrival takes place randomly over 30 years), we have demonstrated the potential impacts of three invertebrates over time. Given these two different model scenarios, there are two different ways we can prioritise species using the model results. Firstly, priorities can be established on the basis of the 'incursion year 1' scenario. This would mean our highest priority species would be wheat stem sawfly (with an annualised total damage cost of $54.5 million), followed by khapra beetle ($46.1 million) and cabbage seedpod weevil ($33.3 million). However, if we take into account the probabilities of each IAS actually making it to WA and becoming established, our highest priority would be cabbage seedpod weevil (with a probability-weighted or expected annual total damage cost of $1.0 million), followed by khapra beetle ($0.9 million) and wheat stem sawfly ($0.7). The change in our priorities caused by including arrival probabilities highlights the difficulty in thinking about and establishing IAS "target lists". While it may be unlikely that the three species we have modelled here will arrive in WA any time soon, the consequences should they do so could be enormous.

4.3 Plant Pathogens

4.3.1 *Outline*

Plant pathogens are disease-causing organisms that attack plants, including viruses, bacteria and fungi. If invertebrate IAS are hard to keep out of a region like WA, plant pathogens are an order of magnitude more difficult. Not only are they smaller and harder to detect, but they can also be extremely hardy and persist in the absence of hosts for long periods of time. Globally, it has been estimated that plant industries experience around 16% lower production as a result of plant pathogens than they otherwise would (Oerke, 2006). Making matters worse is the fact that in many cases we do not know much about the plant pathogens that potentially threaten a region like WA. It is probable that less than 10% of the world's fungi are known to science, and of those very few are known well in terms of their potential pathogenicity (Palm and Rossman, 2003). The following examples demonstrate potential plant pathogen losses to an agricultural region like WA using some relatively well-known species.

4.3.2 *Wheat stem rust, Ug99 (Puccinia graminis f. sp. tritici)*

4.3.2.1 *Description*

In 1999, a new race of the stem rust fungal pathogen emerged in Uganda that was found to be able to overcome 17 existing stem rust resistance genes. This includes *Sr31*, the most widely used and effective source of resistance in bread wheat for the past three decades (Periyannan *et al.*, 2013; Nazari *et al.*, 2009). The new race, initially called TTKSK but now known as Ug99, has spread to Northeast Africa and the Arabian Peninsula and now poses a considerable threat to global wheat production, and consequently global food security (Singh *et al.*, 2011; Stokstad, 2007).

Ug99 has been shown to cause grain yield losses of between 6% and 12% in moderately resistant cultivars in Africa up to 70% for highly susceptible cultivars including Chozi, which carries the *Sr31* gene (Macharia and Wanyera, 2012). *Sr31* has not been used widely in Australia, but it is estimated that if Ug99 becomes established up to 60% of cultivars will become moderately susceptible to susceptible (Malipatil and Plant Health Australia, 2008). Since the initial discovery of the pathogen, two new Ug99 derivatives have been identified which indicate the capacity of the fungus to mutate and adapt to new conditions (Malipatil and Plant Health

Australia, 2008). Future mutations could potentially occur that affect resistance genes in use and under development in Australia.

The resistant genes *Sr22* and *Sr26* (as well as possibly *Sr2, Sr12* and *Sr13*) are assumed to form the basis of a response to an outbreak of Ug99 in WA. Due to the difficulties of eradicating a wind-borne fungus, it is likely that following detection widespread replanting to resistant varieties would be undertaken as the primary means of response. A yield penalty of 5–15% (i.e. Pert(0.05, 0.10, 0.15)) is assumed to accompany the replanting (Cook *et al.*, 2011a), and input costs are expected to rise as fungicide treatments are also used.

4.3.2.2 *Parameters*

See Table 4.4.

Table 4.4. Ug99 model parameters

Description	Nil management
Area currently affected, A^{min} (ha).	0
Cost of eradication, E ($/ha).[a]	Uniform(1.0×10^6, 1.0×10^7)
Demand elasticity, η.[b]	∞
Exponential rate of decline for eradication success probability with respect to area affected, ϕ	Pert(-0.2, -0.15, -0.1)
Increased variable cost of production if eradication fails, V ($/ha).[c]	Discrete[(0.0, 13.5, 27.0)(0.5, 1.0, 1.0)]
Intrinsic rate of population/infection density increase, δ (yr^{-1}).[d]	Pert(0.20, 0.35, 0.50)
Intrinsic rate of infestation/infection area growth, r (yr^{-1}).[d]	Pert(1.0, 1.25, 1.5)
Intrinsic rate of satellite generation per unit of area affected, μ (#/ha).[d]	Pert(1.0×10^{-2}, 2.5×10^{-2}, 5.0×10^{-2})
Maximum area affected, A^{max} (ha).[e]	4.7×10^6
Maximum area considered for eradication, A^{erad} (ha)	Pert(5.0×10^3, 7.5×10^3, 1.5×10^3)
Maximum infestation/infection density, K (#/ha).[d]	Uniform(100, 1000)

(Continued)

Table 4.4.   (*Continued*)

Description	Nil management
Maximum number of satellite sites generated in a single time step, s^{max} (#).[d]	Pert(70, 85, 100)
Minimum infestation/infection density, N^{min}(#/ha).[d]	1.0×10^{-4}
Minimum number of satellite sites generated in a single time step, s^{min} (#).	1.0
Population/infection diffusion coefficient, D (ha/yr).[d]	Pert(4, 6, 8)
Prevailing price for affected commodities in the first time step, P_0 ($/T). [e]	Wheat 280
Probability of detection (%).[f]	Uniform(30, 50)
Probability of entry and establishment, z (%).[f]	Uniform(7.0×10^{-2}, 5.0)
Reduction in export earnings attributable to a loss of pest/disease area freedom, β(%).[f]	Uniform(0, 5)
Yield reduction despite control, Y(%).[f]	Uniform(5, 15)

[a]Specified with reference to Horn and Breeze (2000); [b]Host products are predominantly sold to export market and WA is not sufficiently large to exert pressure on the world price; [c]Propiconazole ($13.50/L) at 250–500 mL/ha applied between 0 and 2 times per season (Rural Solutions SA, 2014); [d]Specified with reference to Waage *et al.* (2005) and Cook *et al.* (2011a); [e]ABS (2014) and Malipatil and Plant Health Australia (2008); [f]Cook *et al.* (2011a).

4.3.2.3 *Results*

See Figures 4.9–4.11.

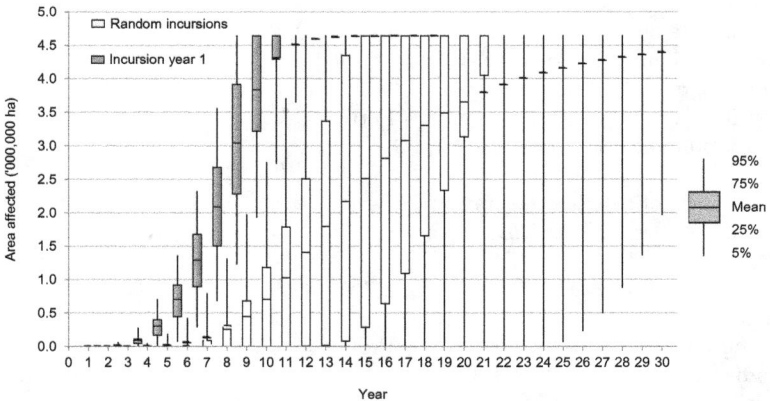

Figure 4.9.   Area affected by Ug99 in WA if an incursion occurs in year 1 and if incursions occur randomly over 30 years

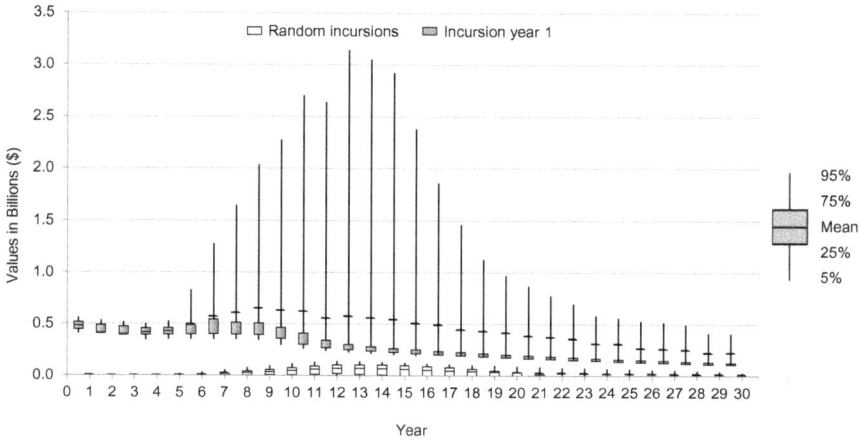

Figure 4.10. Total damage cost of Ug99 in WA if an incursion occurs in year 1 and if incursions occur randomly over 30 years

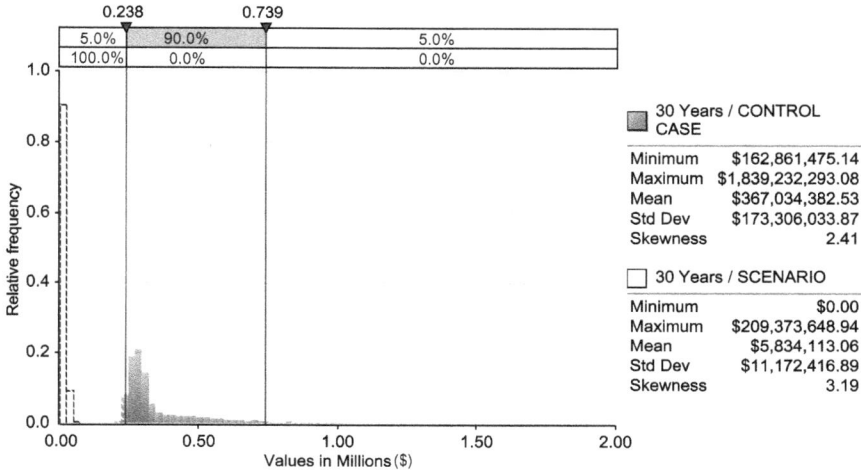

Figure 4.11. Annualised total damage cost of Ug99 in WA if an incursion occurs in year 1 and if incursions occur randomly over 30 years

4.3.2.4 *Conclusion*

There is a lot of uncertainty about the impact of Ug99 if it were to enter WA and become established. Although resistant varieties are grown in the State, the yield reduction associated with growers having to use highly resistant varieties as the fungus spreads is difficult to predict with certainty. Moreover, the demonstrated capacity of Ug99 to evolve and overcome resistance means the impact of the fungus is even more difficult to predict. Hence, although relatively minor impacts are possible, average total damage costs of over $0.4 billion per year are estimated if an incursion were to occur in year 1. Random incursions over a 30-year period are expected to cause average total damage costs of almost $27.7 million per year.

4.3.3 *Barley stripe rust* (*Puccinia striiformis* f. sp. *hordei*)

4.3.3.1 *Description*

Barley stripe rust (*Puccinia striiformis* f. sp. *hordei*) fungus has been shown to affect approximately 80% of current cultivars and advanced breeding lines present in Australia, and as such presents a serious threat to WA grains industries (Wellings *et al.*, 2000; Wellings and Plant Health Australia, 2010). The rust is not present in Australia, but has caused serious problems to winter barley production in Europe, South America more recently North America. Since 1991, it has quickly spread and become established in the South-Central and Western United States and is now the most important disease of barley in Western United States (Line, 2002).

Barley stripe rust is spread via wind and primarily attacks plant leaves. In heavy infestations leaf sheaths and heads can also be affected, but it is not seed-borne. Hence, it is not expected to cause export losses. The primary symptom of stripe rust is the appearance of yellow or orange pustules arranged in stripes along upper leaf surfaces; hence its common name. Infection reduces plant vigour and root growth, increases water loss and hinders photosynthesis. This can result in yield losses of up to 70%, depending on the growth stage of the crop when the rust develops. Early infections are the most damaging (Wellings and Plant Health Australia, 2010).

Given the frequency of travel between Australia and countries where the pathogen exists, together with the ability of fungal spores to be carried on contaminated clothing for considerable periods of time, likelihood of barley stripe rust entering WA is considered relatively high (Wellings *et al.*, 1987). Eradication is highly unlikely, and would only be technically feasible if the rust was detected while still contained within a small area and the spore load is light. If this is not the case and the rust spreads throughout the State, barley crops would need to be treated with additional fungicides in order to remain economically viable. This will add substantially to production costs, but will reduce yield losses to the disease substantially.

4.3.3.2 *Parameters*

See Table 4.5.

Table 4.5. Barley stripe rust model parameters

Description	Nil management
Area currently affected, $A^{\min}$ (ha).	0
Cost of eradication, E ($/ha).[a]	Uniform(1.0×10^5, 1.0×10^6)
Demand elasticity, η.[b]	∞
Exponential rate of decline for eradication success probability with respect to area affected, ϕ	Pert(-0.2, -0.15, -0.1)
Increased variable cost of production if eradication fails, V ($/ha).[c]	Discrete[(0.0, 13.5, 27.0)(0.5, 1.0, 1.0)]
Intrinsic rate of population/infection density increase, δ (yr^{-1}).[d]	Pert(0.20, 0.35, 0.50)
Intrinsic rate of infestation/infection area growth, r (yr^{-1}).[d]	Pert(1.0, 1.25, 1.5)
Intrinsic rate of satellite generation per unit of area of affected, μ (#/ha).[d]	Pert(1.0×10^{-2}, 2.5×10^{-2}, 5.0×10^{-2})
Maximum area affected, $A^{\max}$ (ha).[e]	1.1×10^6
Maximum area considered for eradication, A^{erad} (ha)	Uniform(5.0×10^2, 1.5×10^3)
Maximum infestation/infection density, K (#/ha).[d]	Uniform(100, 1000)
Maximum number of satellite sites generated in a single time step, $s^{\max}$ (#).[d]	Pert(70, 85, 100)

(Continued)

Table 4.5. (*Continued*)

Description	Nil management
Minimum infestation/infection density, $N^{\min}$ (#/ha).[d]	1.0×10^{-4}
Minimum number of satellite sites generated in a single time step, $s^{\min}$ (#).	1.0
Population/infection diffusion coefficient, D (ha/yr).[d]	Pert(4, 6, 8)
Prevailing price for affected commodities in the first time step, P_0 ($/T).[e]	Barley 213
Probability of detection (%).[d]	Uniform(10, 40)
Probability of entry and establishment, z (%).[f]	Uniform(1.5, 21.0)
Reduction in export earnings attributable to a loss of pest/disease area freedom, β (%).[f]	Uniform(0, 5)
Yield reduction despite control, Y (%).[f]	Uniform(5, 15)

[a]Specified with reference to Horn and Breeze (2000); [b]Host products are predominantly sold to export market and WA is not sufficiently large to exert pressure on the world price; [c]Propiconazole ($13.50/L) at 250–500 mL/ha applied between 0 and 2 times per season (Rural Solutions SA, 2014); [d]Specified with reference to Waage *et al.* (2005); [e]ABS (2014); [f]Wellings and Plant Health Australia (2010).

4.3.3.3 *Results*

See Figures 4.12–4.14.

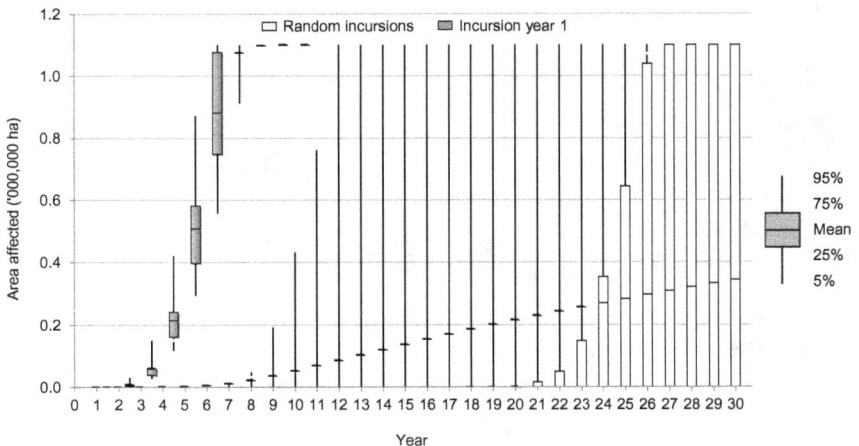

Figure 4.12. Area affected by barley stripe rust in WA if an incursion occurs in year 1 and if incursions occur randomly over 30 years

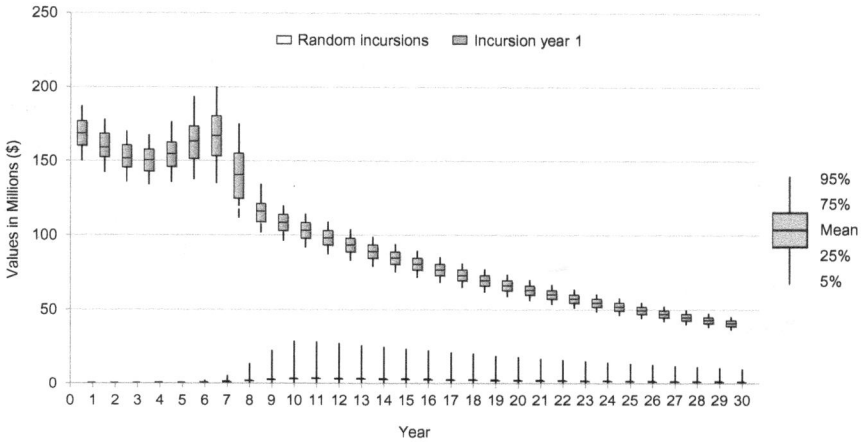

Figure 4.13. Total damage cost of barley stripe rust in WA if an incursion occurs in year 1 and if incursions occur randomly over 30 years

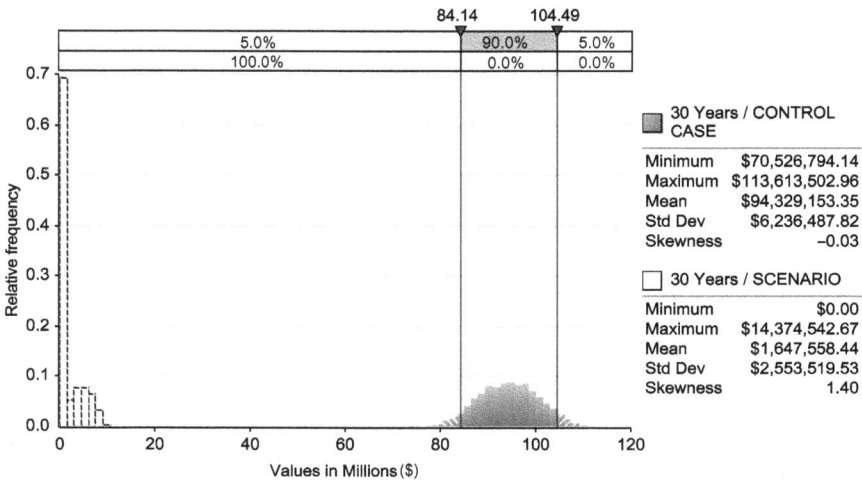

Figure 4.14. Annualised total damage cost of barley stripe rust in WA if an incursion occurs in year 1 and if incursions occur randomly over 30 years

4.3.3.4 *Conclusion*

Barley stripe rust presents a serious threat to WA barley growers due to the relative ease with which it could enter the State, become established and spread. If an incursion were to occur in the current year, it is predicted annual damages of over $94.3 million would be expected over the subsequent 30-year period. These high damage costs are attributable to an extremely rapid rate of spread, yield losses of up to 15% and additional applications of fungicide. Taking the probabilities of barley stripe rust entry and establishment into account, annual damages of $1.6 million are predicted by the model.

4.3.4 *Karnal bunt (Tilletia indica)*

4.3.4.1 *Description*

Karnal bunt (*Tilletia indica*) is a disease of bread wheat, durum wheat and triticale. It is caused by the smut fungus *T. indica* Mitra (= *Neovossia indica* (Mitra) Mundkur) and was first discovered in 1930 at the Botanical Research Station, Karnal, Haryana, in Northwest India (Bonde *et al.*, 1997); hence the name Karnal bunt. It is now found throughout Afghanistan, Iran, Iraq, Lebanon, Nepal and Pakistan (Centre for Agricultural Bioscience International, 2014). It has also been reported in Mexico in 1972, the United States in 1996 and South Africa in 2000 (Stansbury and Pretorius, 2001).

Karnal bunt seldom causes serious yield losses in wheat, but is considered a major threat to wheat exports. The effect of the disease if more than 3% of grains are bunted is to produces trimethylamine (Sharma *et al.*, 2012). This is non-toxic to humans, but emits a fishy odour and causes discolouration.

Contaminated seed is considered a major source of disease spread and since 1983 Karnal bunt has been considered serious enough to warrant the imposition of planting and seed industry quarantines and restrictions. Price reductions and outright rejection of infected seed is to be expected if the pathogen were to become established in WA. This may be equivalent to an export loss of around 30% (Brennan *et al.*, 1992). The impact could therefore be extreme despite the use of fungicides to suppress the natural spread of the disease in affected areas.

However, due to the trade restrictions on imported grains, the likelihood of WA experiencing a Karnal bunt outbreak is small. Stansbury and Pretorius (2001) conclude that an entry event can be expected one year in 25, and an establishment event one year in 67. Hence, in the analysis to follow there is a large disparity in the impacts of the disease between the incursion year 1 and random incursion scenarios.

4.3.4.2 *Parameters*

See Table 4.6.

Table 4.6. Karnal bunt model parameters

Description	Nil management
Area currently affected, $A^{\min}$ (ha).	0
Cost of eradication, E (\$/ha).[a]	Uniform$(1.0 \times 10^6, 1.0 \times 10^7)$
Demand elasticity, η.[b]	∞
Exponential rate of decline for eradication success probability with respect to area affected, ϕ	Pert$(-0.2, -0.15, -0.1)$
Increased variable cost of production if eradication fails, V (\$/ha).[c]	Discrete$[(0.0, 13.5, 27.0)(0.5, 1.0, 1.0)]$
Intrinsic rate of population/infection density increase, δ (yr^{-1}).[d]	Pert$(0.20, 0.35, 0.50)$
Intrinsic rate of infestation/infection area growth, r (yr^{-1}).[d]	Pert$(1, 2, 3)$
Intrinsic rate of satellite generation per unit of area affected, μ(#/ha).[d]	Pert$(1.0 \times 10^{-2}, 2.5 \times 10^{-2}, 5.0 \times 10^{-2})$
Maximum area affected, $A^{\max}$ (ha).[e]	4.7×10^6
Maximum area considered for eradication, A^{erad} (ha).	Pert$(5.0 \times 10^3, 7.5 \times 10^3, 1.5 \times 10^4)$
Maximum infestation/infection density, K (#/ha).[d]	Uniform$(1.0 \times 10^4, 1.0 \times 10^5)$
Maximum number of satellite sites generated in a single time step, $s^{\max}$ (#).[d]	Pert$(10, 15, 20)$
Minimum infestation/infection density, $N^{\min}$ (#/ha).[d]	Pert$(0.00, 0.025, 0.050)$

(Continued)

Table 4.6. (*Continued*)

Description	Nil management
Minimum number of satellite sites generated in a single time step, s^{min} (#).	1.0
Population/infection diffusion coefficient, D (ha/yr).[d]	Pert(4, 6, 8)
Prevailing price for affected commodities in the first time step, P_0 (\$/T).[e]	Wheat 280
Probability of detection (%).	Pert(20, 40, 60)
Probability of entry and establishment, z (%).[f]	Uniform(1.0×10^{-4}, 2.5×10^{-2})
Reduction in export earnings attributable to a loss of pest/disease area freedom, β.[g]	Uniform(30, 50)
Yield reduction despite control, Y(%).[g]	Uniform(0, 1)

[a]Specified with reference to Horn and Breeze (2000); [b]Host products are predominantly sold to export market and WA is not sufficiently large to exert pressure on the world price; [c]Propiconazole (\$13.50/L) at 250–500 mL/ha applied between 0 and 2 times per season (Rural Solutions SA, 2014); [d]Specified with reference to Waage *et al.* (2005); [e]ABS (2014); [f]Stansbury and Pretorius (2001); [g]Murray and Brenan (1998).

4.3.4.3 *Results*

See Figures 4.15–4.17.

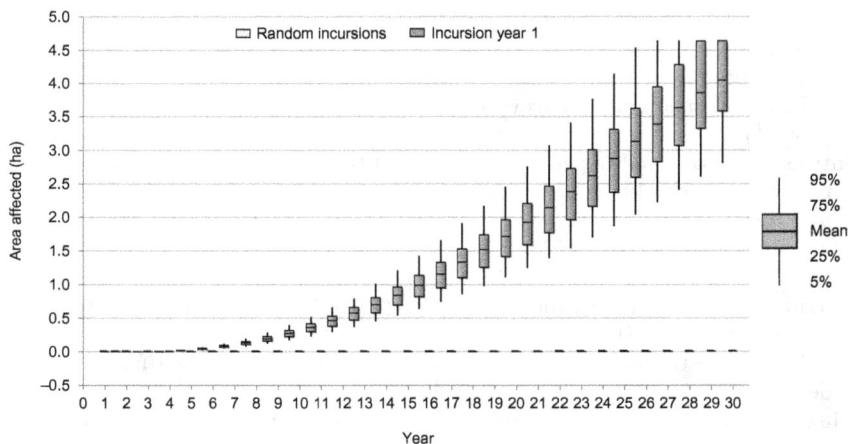

Figure 4.15. Area affected by Karnal bunt in WA if an incursion occurs in year 1 and if incursions occur randomly over 30 years

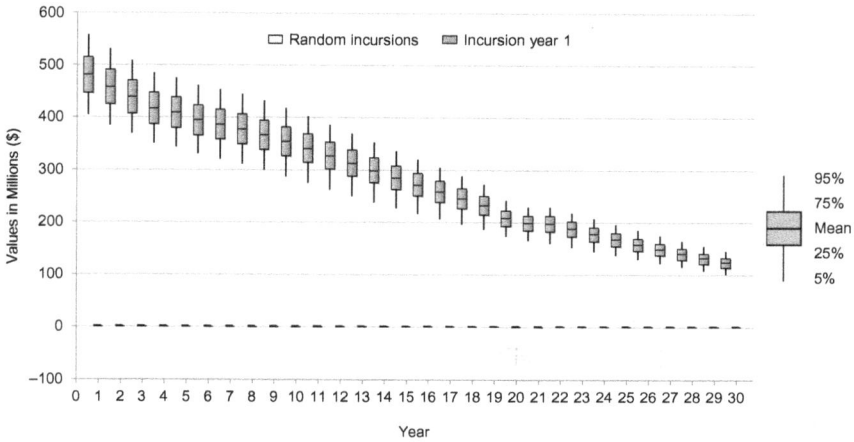

Figure 4.16. Total damage cost of Karnal bunt in WA if an incursion occurs in year 1 and if incursions occur randomly over 30 years

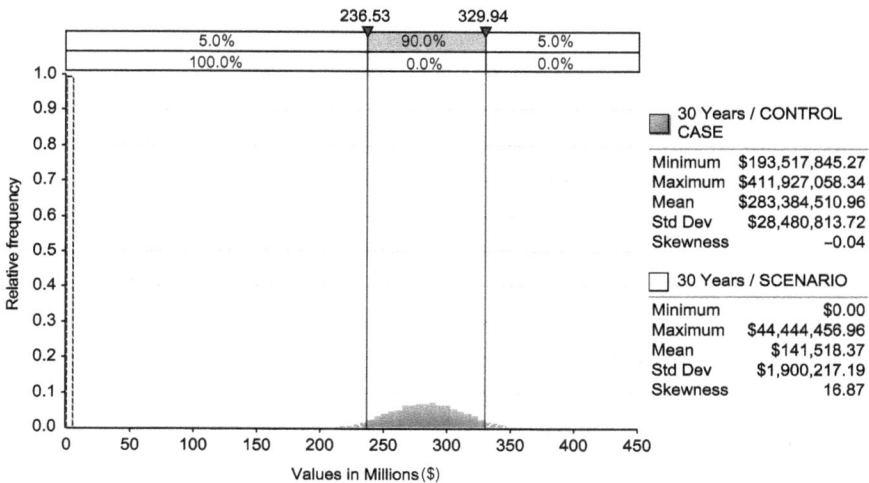

Figure 4.17. Annualised total damage cost of Karnal bunt in WA if an incursion occurs in year 1 and if incursions occur randomly over 30 years

4.3.4.4 *Conclusion*

If Karnal bunt were to be present in WA in the current time period the resultant total damage costs would be extremely large due to the loss of export markets. Over the subsequent 30 years, these losses are expected to average close to $0.3 billion per year. However, given that it is highly unlikely the disease will enter and spread within the State, the total damage costs expected under the random incursion scenario are only $0.1 million per year. This highlights the difficulties in perceiving and prioritising rare events with extreme consequences. The likelihood that WA will face a Karnal bunt incursion scenario is remote, but if and when it occurs the effects on the grains sector could be devastating.

4.3.5 *Summary — plant pathogens*

Plant pathogens represent a very challenging taxonomic group to work with from a biosecurity policy perspective as there are many uncertainties about their likelihood of arrival in a region like WA and their impact if and when they arrive. If we were to prioritise the three case studies we have looked at in this section on the basis of their expected costs in an 'incursion year 1' scenario, our highest priority species would be Ug99 (with an annualised total damage cost of $0.4 billion), followed by Karnal bunt ($0.3 billion) and barley stem rust ($94.3 million). If we then factor in the probabilities of each species arriving in WA and establishing, our highest priority remains Ug99 (with an expected annual total damage cost of $27.7 million), followed by barley stem rust ($1.6 million) and Karnal bunt ($0.1). The prominence of Karnal bunt in particular is severely affected by its low probability of arrival, but if any of the three species do become established in WA our results suggest they could have a tremendous impact on host industries.

4.4 Vertebrates

4.4.1 *Outline*

Over 80 vertebrate animal species have been introduced to and become established in Australia since the first fleet of European settlers arrived in 1788 (Bomford and Hart, 2002), and at least one species prior to this (Oskarsson *et al.*, 2011). Almost half are now considered pests. Despite being much larger than invertebrate or pathogenic plant pests, vertebrates too can be extremely difficult to exclude from regions they are well adapted to. Bird species, for instance, can travel vast distances unimpeded by biosecurity controls. Problem species also tend to have a broad diet, enabling them to survive and flourish in both farming areas and in natural ecosystems. In both environments their impacts can be severe, but we will focus on agricultural impacts in this section.

Given the number of vertebrate IAS established in WA it is prudent to include species that have already been introduced in our examples, but we also include a species yet to be introduced to the State (but present in other States of Australia) and a native Eastern-Australian species that has been introduced to WA. The two scenarios we simulate with our model for each case study include one in which biosecurity effort is expended to mitigate IAS damage and another where it is not (i.e. nil management). Where possible, existing response costs have been included to provide indicative returns on investment in activities targeted at specific species. Also note that instead of providing box whisker plots of predicted area occupied by each species over time, as in Sections 4.2 and 4.3 (and 4.5 to follow) we instead provide population estimates.

4.4.2 *Wild rabbit (Oryctolagus cuniculus)*

4.4.2.1 *Description*

Rabbits have been present in Australia for more than two centuries and have become one of the country's most serious introduced vertebrate pest problems (Thompson and King, 1994). Rabbits cause considerable damage to agricultural production, ranging from direct competition for forage with sheep and cattle to damage to yield effects on high value horticultural crops (Scanlan *et al.*, 2006). Hence, their control on agricultural land is important and requires the integration of biological, chemical and mechanical methods.

Biological control for rabbits, including the myxoma virus causing the disease myxomatosis, has been particularly effective. Released in 1950, the virus initially killed over 90% of rabbits that were exposed to it. Despite resistance becoming an issue over time, myxomatosis persists in the population and maintain rabbit numbers at less than 25% of their former abundance (SEWPaC, 2011). The other important biological control is rabbit haemorrhagic disease virus (RHDV). This disease has proved more effective in wetter parts of the country than in drier regions. Australian rabbits have recently been discovered to carry a native calicivirus that may confer some immunity to RHDV, and research is ongoing to identify more effective field strains to release in Australia (Strive *et al.*, 2010). The main chemical control used for rabbits is the poison sodium flouroacetate (1080), although pressure fumigation or diffusion fumigation using toxins like chloropicrin and carbon monoxide are also used to kill rabbits within warrens. Mechanical control methods involve the destruction (or "ripping") of warrens and shooting.

De Milliano *et al.* (2010) revealed a number of gaps at the State and regional levels affecting, amongst other activities, biosecurity activities related to rabbit control. These included a lack of regional IAS management plan. It was also noted that while a small number of specific management strategies existed at the time for individual IAS (e.g. wild dogs, starlings, cane toads and rainbow lorikeets), most IAS lacked a formal management strategy (De Milliano *et al.*, 2010). As a consequence, the control of rabbits lacked coordination at a regional level.

Currently, a public and private cost sharing arrangement applies to a range of vertebrate IAS in WA at different spatial scales (detailed in Box 13). Under the terms and conditions of the *Biosecurity and Agriculture Management Act 2007* (Parliament of Western Australia, 2007), this involves Declared Species Groups (DSGs), Recognised Biosecurity Groups (RBGs) and Industry Funding Schemes (IFSs). These arrangements have not led to sizeable pressure being applied to the wild rabbit population.

Box 13. Cost Sharing Arrangements for Wild Rabbit Control in Western Australia

A multi-faceted public and private cost sharing arrangement applies to a range of vertebrate pests in WA, including wild rabbits. Under the terms and conditions of the *Biosecurity and Agriculture Management Act 2007* (Parliament of Western Australia, 2007), this involves Declared Species Groups (DSGs), Recognised Biosecurity Groups (RBGs) and Industry Funding Schemes (IFSs):

1. DSGs are temporary, non-statutory action groups formed to manage a single pest species at a localised scale. They are funded via voluntary contributions that are matched dollar-for-dollar by State government over a specified period of time. There are currently DSGs that have been formed to manage wild dogs (*Canis lupus dingo, C. l. familiaris* and *C. l. dingo × C. l. familiaris*), rainbow lorikeets (*Trichoglossus haematodus*) and wild boar (*Sus scrofa*);

2. RBGs are statutory groups formed to manage multiple IAS with impacts at a regional scale. Through the *Biosecurity and Agriculture Management Act 2007*, organised land management groups that control declared species can receive formal recognition as RBGs. This status entitles the group to receive funding for IAS management activities raised from Agriculture Protection Rates (APR) applied to all land within the State held under Crown Pastoral Leases. APR funds are matched dollar-for-dollar by the State Government and the combined funding is used to undertake a range of IAS control programs. There are currently five RBGs established in the State, three of which (Goldfields Nullarbor, Meekatharra and Carnarvon) have rabbit populations. Although these groups could carry out control operations or engaging contractors to manage rabbits, there are currently no such programs in place;

3. IFSs are statutory action groups established to govern the management of multiple IAS affecting a given industry. There are currently three IFSs operating in the State for (i) cattle, (ii) grains, seeds and hay and (iii) sheep and goats.

It follows that there are minimal differences in terms of parameters used to characterise the *current management* and *nil management* scenarios in this assessment. The current management strategy involves DAFWA disseminating information and advice about the control of rabbits on agricultural land, which in turn increases the effectiveness of controls put in place by land managers. Direct action to remove rabbits will not generally occur.

4.4.2.2 *Parameters*

See Table 4.7.

4.4.2.3 *Results*

See Figures 4.18–4.20.

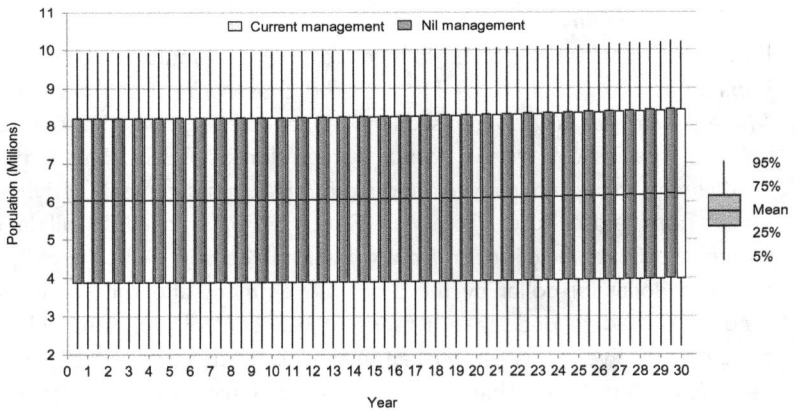

Figure 4.18. Expected rabbit population in WA under current management and nil management scenarios

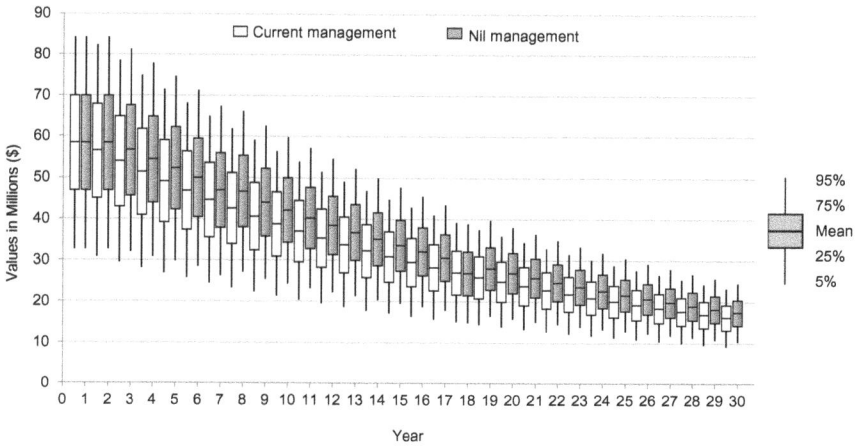

Figure 4.19. Predicted industry losses from rabbits in WA under the current management and nil management scenarios

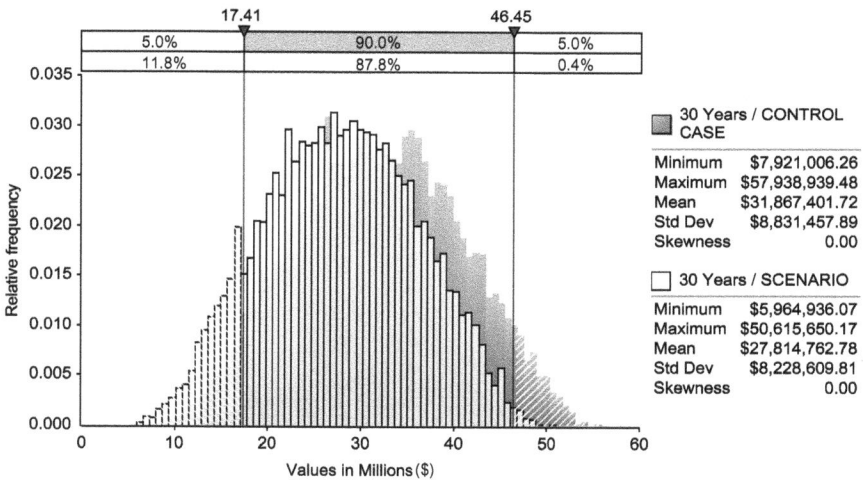

Figure 4.20. Annualised average producer cost from rabbits in WA over a 30-year period under the current management scenario and a nil management scenario

Table 4.7. Rabbit model parameters

Description	Nil management				Current management			
Area currently affected, A^{min} (ha). [a]	3.8×10^7				3.8×10^7			
Cost of eradication, E ($/ha). [a]	na				na			
Demand elasticity, η. [b]	Uniform(−1.5, −1.3)				Uniform(−1.5, −1.3)			
Exponential rate of decline for eradication success probability with respect to area affected, ϕ	na				na			
Increased variable cost of production if eradication fails,	Barley	Uniform(0, 1)	Oats	Uniform(0, 1)	Barley	Uniform(0, 0.5)	Oats	Uniform(0, 0.5)
	Beef	Uniform(0, 2)	Wine grapes	Uniform(0, 60)	Beef	Uniform(0, 1)	Wine grapes	Uniform(0, 30)
V ($/ha). [c]	Broccoli	Uniform(0, 10)	Wool	Uniform(0, 1)	Broccoli	Uniform(0, 5)	Wool	Uniform(0, 1)
	Lettuce	Uniform(0, 2)	Wheat	Uniform(0, 1)	Lettuce	Uniform(0, 1)	Wheat	Uniform(0, 0.5)
Intrinsic rate of population density increase, δ (yr^{-1}). [d]	Uniform(0.05, 0.10)				Uniform(0.05, 0.10)			
Intrinsic rate of infestation area growth, r (yr^{-1}). [e]	Pert(1.00, 1.25, 1.50)				Pert(1.00, 1.25, 1.50)			

Parameter		
Intrinsic rate of satellite generation per unit area of infestation, μ (#/ha).[d]	Pert(5.0×10^{-2}, 7.5×10^{-2}, 1.0×10^{-1})	Pert(5.0×10^{-2}, 7.5×10^{-2}, 1.0×10^{-1})
Maximum area affected, A^{max} (ha). [a]	4.9×10^{7}	4.9×10^{7}
Maximum area considered for eradication, A^{erad} (ha)	na	na
Maximum infestation density, K (#/ha).[d]	Uniform(5, 10)	Uniform(5, 10)
Maximum number of satellite sites generated in a single time step, s^{max} (#).[d]	Pert(10, 15, 20)	Pert(10, 15, 20)
Minimum infestation density, N^{min} (#/ha).[d]	Pert(0.00, 0.025, 0.050)	Pert(0.00, 0.025, 0.050)
Minimum number of satellite sites generated in a single time step, s^{min} (#).	1.0	1.0

(Continued)

Table 4.7. *(Continued)*

Description	Nil management		Current management	
Population diffusion coefficient, D (ha/yr).[d]	Pert(2, 3, 4)		Pert(2, 3, 4)	
Prevailing price for affected commodities in the first time step, P_0 ($/T).[e]	Barley	213	Barley	213
	Oats	176	Oats	176
	Beef grapes	1730	Beef	1730
	Wine	1252	Wine grapes	1252
	Broccoli	7960	Broccoli	7960
	Wool	10710	Wool	10710
	Lettuce	2620	Lettuce	2620
	Wheat	280	Wheat	280
Probability of re-entry and establishment, z (%).	1.0		1.0	
Reinfestation detection probability.	na		na	
Yield reduction despite control, Y (%).[f]	Barley	Uniform(0, 1)	Barley	Uniform(0, 0.5)
	Oats	Uniform(0, 1)	Oats	Uniform(0, 0.5)
	Beef	Uniform(0, 2)	Beef	Uniform(0, 1)
	Wine grapes	Uniform(0, 2)	Wine grapes	Uniform(0, 1)
	Broccoli	Uniform(0, 2)	Broccoli	Uniform(0, 2.5)
	Wool	Uniform(0, 5)	Wool	Uniform(0, 1)
	Lettuce	Uniform(0, 1)	Lettuce	Uniform(0, 2.5)
	Wheat	Uniform(0, 5)	Wheat	Uniform(0, 0.5)

[a] ABS (2012) and ABARES (2011); [b]Ulubasoglu et al. (2011); [c]Hunter et al. (2008); [d]Waage et al. (2005); [e]ABS (2012) and Curtis and McCormick (2012); [f]Based in part on Gong et al. (2009).

4.4.2.4 *Conclusion*

Due to the effects of myxomatosis and RHDV, rabbit numbers in WA are substantially below historical levels. However, they are still capable of producing large losses. Under a nil management scenario it is estimated average damages would exceed $31.9 million per annum over 30 years. As rabbits occupy multiple regions and affect multiple industries, they are difficult to manage under the current DSG, IFS and RBG cost sharing arrangements. Consequently, pressure to the population in addition to that exerted by biological control agents is not being applied. It is estimated that current management activities only produce around $4.1 million of agricultural benefits per year.

4.4.3 *European starling (Sturnus vulgaris)*

4.4.3.1 *Description*

The European starling (*Sturnus vulgaris*), henceforth starling, origi-nated in Europe and Asia and has since become established in North America, South Africa, New Zealand and Australia (Linz *et al.*, 2007). The species was first introduced into Australia in the mid-19th century and are now widespread in all States with the exception of WA. Starling populations were first sighted in WA as early as 1936, but they were not actively managed until the early 1970s (Woolnough *et al.*, 2005).

Today, starlings are all but absent from the State, but periodi-cally birds cross the Nullarbor Plain and enter WA near the border with South Australia or from other unknown areas. In response, detection and control campaigns since the 1970s have succeeded in local eradication of birds from specific areas and have kept numbers of birds down throughout the south coast region of WA (Woolnough *et al.*, 2005).

Through their feeding habits starlings have an adverse impact on beef, canola, cereals, citrus fruit, grapes, pome fruit and stone fruit production. This analysis compares damages that would result in these industries under two scenarios. Firstly, a nil management approach is considered in which starlings would be permitted to enter and become established in the State. The second scenario involves a sustained *ongoing control* policy. It is assumed that this would involve an investment of \$1.2 million per annum (i.e. 2009-level funding scenario outlined in Campbell and Wilkinson, 2014).

4.4.3.2 *Parameters*

See Table 4.8.

4.4.3.3 *Results*

See Figures 4.21–4.23.

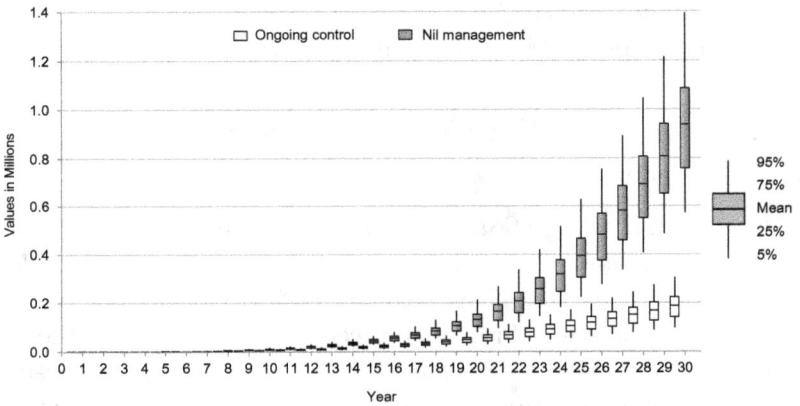

Figure 4.21. Expected population of starlings in WA under the ongoing control and nil management scenarios

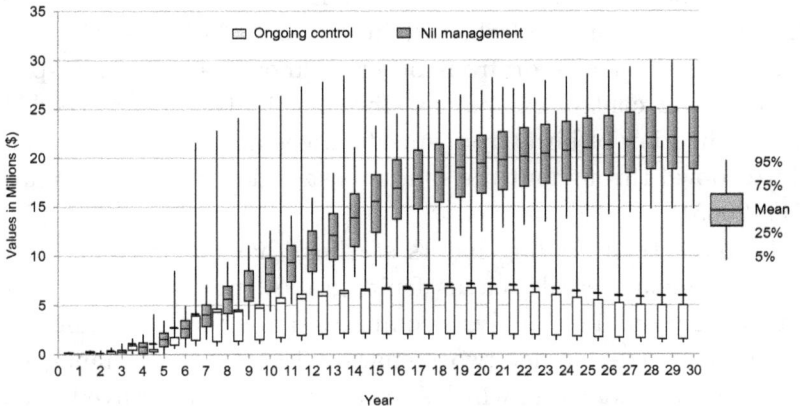

Figure 4.22. Predicted industry losses from starlings in WA under the ongoing control and nil management scenarios

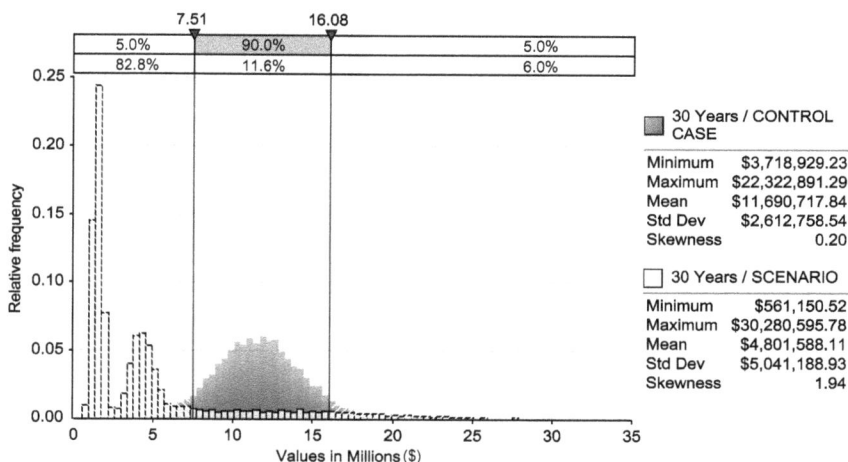

Figure 4.23. Annualised average producer cost from starlings in WA over a 30-year period under the ongoing control scenario and a nil management scenario

4.4.3.4 *Conclusion*

Starlings present a large threat to WA agriculture. They have the potential to affect a broad variety of crops over a large area and have a proven capacity to enter WA and become established. This analysis predicts that the impact of the starlings on beef, canola, cereals, citrus fruit, grapes, pome fruit and stone fruit production, particularly in the long term, is sufficiently large to warrant an investment of up to $6.9 million per year (based on a 30-year average estimate). This would indicate that the investment of funds in 2009 (see Campbell and Wilkinson (2014)), for example, was conservative despite being greater than the current investment. If no action is taken to prevent the establishment of starlings in WA annual damages are expected to average $11.7 million over the next 30 years.

4.4.4 *Rainbow lorikeet (Trichoglossus haematodus)*

4.4.4.1 *Description*

Rainbow lorikeets are a native species of Eastern Australia, Indonesia, New Guinea, the Solomon Islands, Vanuatu and New Caledonia (Chapman, 2005). They were introduced to WA in the 1960s and

Table 4.8. Starling model parameters

Description	Nil management	Ongoing control
Area currently affected, $A^{\min}$ (ha).[a]	Uniform(0,100)	0
Cost of eradication, E ($/ha).[a]	0.0	Pert(2.0×10^5, 5.0×10^5, 7.5×10^5)
Demand elasticity, η.[b]	Uniform(−1.5, −1.3)	Uniform(−1.5, −1.3)
Exponential rate of decline for eradication success probability with respect to area affected, ϕ	Pert(−0.1, −0.15, −0.2)	Pert(−0.1, −0.15, −0.2)
Increased variable cost of production if eradication fails, V ($/ha).[c]	Discrete([0, 106, 212],[2.5, 0.5, 0.25])	Discrete([0, 106, 212],[1, 0.5, 0.25])
Intrinsic rate of population density increase, δ (yr^{-1}).[d]	Uniform(0.05, 0.10)	Uniform(0.05, 0.10)
Intrinsic rate of infestation area growth, r (yr^{-1}).[e]	Pert(0.020, 0.145, 0.270)	Pert(0.020, 0.145, 0.270)
Intrinsic rate of satellite generation per unit area of infestation, μ (#/ha).[d]	Pert(5.0×10^{-2}, 7.5×10^{-2}, 1.0×10^{-1})	Pert(5.0×10^{-2}, 7.5×10^{-2}, 1.0×10^{-1})
Maximum area affected, $A^{\max}$ (ha).[a]	7.6×10^6	7.6×10^6
Maximum area considered for eradication, A^{erad} (ha)	0.0	10,000
Maximum infestation density, K (#/ha).[d]	Uniform(5,10)	Uniform(5, 10)
Maximum number of satellite sites generated in a single time step, $s^{\max}$ (#).[e]	Pert(10, 15, 20)	Pert(10, 15, 20)
Minimum infestation density, $N^{\min}$ (#/ha).[e]	Pert(0.00, 0.025, 0.050)	Pert(0.00, 0.025, 0.050)
Minimum number of satellite sites generated in a single time step, $s^{\min}$ (#).	1.0	1.0
Population diffusion coefficient, D (ha/yr).[d]	Pert(1.0, 1.5, 2.0)	Pert(1.0, 1.5, 2.0)

Parameter	Scenario 1		Scenario 2	
Prevailing price for affected commodities in the first time step, P_0 ($/T).[f]	Barley 213 Beef 1730 Berries 6284 Canola 542 Citrus 1554 Oats 176	Pome fruit 1193 Stone fruit 2148 Sultanas 3182 Table grapes 3378 Wheat 280 Wine grapes 1252	Barley 213 Beef 1730 Berries 6284 Canola 542 Citrus 1554 Oats 176	Pome fruit 1193 Stone fruit 2148 Sultanas 3182 Table grapes 3378 Wheat 280 Wine grapes 1252
Probability of re-entry and establishment, z (%).	1.0		Uniform(20, 70)	
Reinfestation detection probability.	0.0		Binomial(1.0, 0.6)	
Yield reduction despite control, Y (%).[g]	Barley Uniform(0, 2) Beef Uniform(0,5) Berries Uniform(5, 15) Canola Uniform(0, 2) Citrus Uniform(5, 15) Oats Uniform(0, 2)	Pome fruit Uniform(5,15) Stone fruit Uniform(5, 15) Sultanas Uniform(5, 15) Table grapes Uniform(5, 15) Wheat Uniform(0, 2) Wine grapes Uniform(5, 15)	Barley Uniform(0, 2) Beef Uniform(0, 5) Berries Uniform(5, 15) Canola Uniform(0, 2) Citrus Uniform(5, 15) Oats Uniform(0, 2)	Pome fruit Uniform(5, 15) Stone fruit Uniform(5, 15) Sultanas Uniform(5, 15) Table grapes Uniform(5, 15) Wheat Uniform(0, 2) Wine grapes Uniform(5, 15)

[a] Based on Campbell and Wilkinson (2014) '2009' scenario; [b] Ulubasoglu *et al.* (2011); [c] Assumptions: (i) labour costs of $100/ha (i.e. 3 hr/ha), (ii) 10 shotgun cartridges per ha at $0.30/per cartridge (Tracey *et al.*, 2007), (iii) firearm maintenance costs of $3/ha, and (iv) it is 5 times more likely that 0 culls will be conducted an affected hectare of land by private land managers than 1 cull, and 10 times more likely than 2 culls; [d] Specified with reference to Paton *et al.* (2005); [e] Waage *et al.* (2005); [f] ABS (2012) and Curtis and McCormick (2012); [g] Linz *et al.* (2007).

from an estimated 10 birds initially released the population in metropolitan Perth and across the Southwest of the State was estimated to have reached 20,000 by the early 2000s (Chapman and Massam, 2007). While their numbers have mainly been concentrated in the Perth metropolitan area, rainbow lorikeets could potentially occupy vast areas of the State. Indeed, since 1990 they have been recorded in Esperance, Mandurah, Coolgardie, Boddington, Boyup Brook, Bunbury, Carnamah, Katanning, Williams, Geraldton, Bailingup, Bridgetown and on Rottnest Island (Chapman, 2005).

In response to community concern about the growing WA lorikeet population, the *Rainbow Lorikeet Working Group WA* was established in 2004. This group initially consisted of representatives from the Agriculture Protection Board (abolished in 2010), the Department of Environment and Conservation (now the Department of Parks and Wildlife), DAFWA, the Western Australian Museum, a former member of State Parliament, the Department of Local Government and Regional Development, Birds Australia Western Australia, the Westralia Airports Corporation, the United Bird Societies of WA, the City of Swan, Wine Industry Association and the Grape Growers Association. The Working Group prepared a Rainbow Lorikeet Management Strategy which proposed the removal of at least 4,000 lorikeets from the Perth population each year from 2008, with follow-up activities after seven years to remove at least 500 birds per year to maintain the population at approximately 1,000 birds (Rainbow Lorikeet Working Group, 2008).

This analysis compares a continuation of this strategy (termed *current management*) against a nil management scenario, and only considers the impacts on agricultural production. Elsewhere in Australia, lorikeets are considered a serious pest of pome fruit, stone fruit, citrus, berries, grapes and nuts (Chapman, 2005; Tracey *et al.*, 2007). Although damage to a wide range of other crops is possible, this analysis is limited to potential impacts on fruit. Results should therefore be treated as conservative damage estimates.[3]

---

[3]Lorikeets also pose non-market risks to WA native fauna, and social risks in terms of excessive noise and damage to property and dwellings. They can transfer pathogens such as Psittacine beak and feather disease to native parrots of the region, and also displace

4.4.4.2 *Parameters*

See Table 4.9.

4.4.4.3 *Results*

See Figures 4.24–4.26.

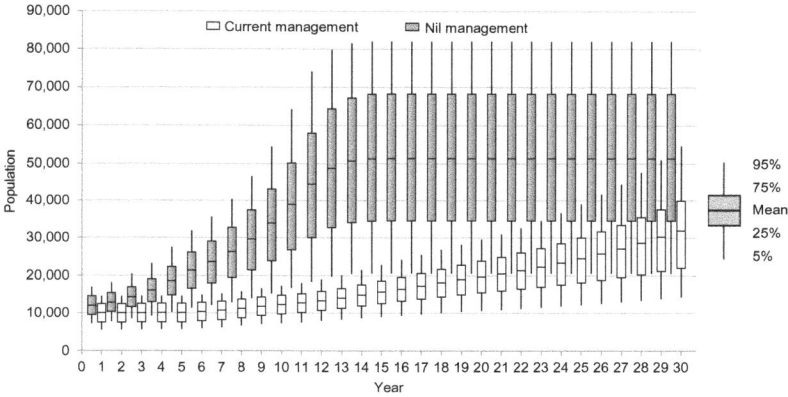

Figure 4.24. Expected rainbow lorikeet population in WA under current management and nil management scenarios

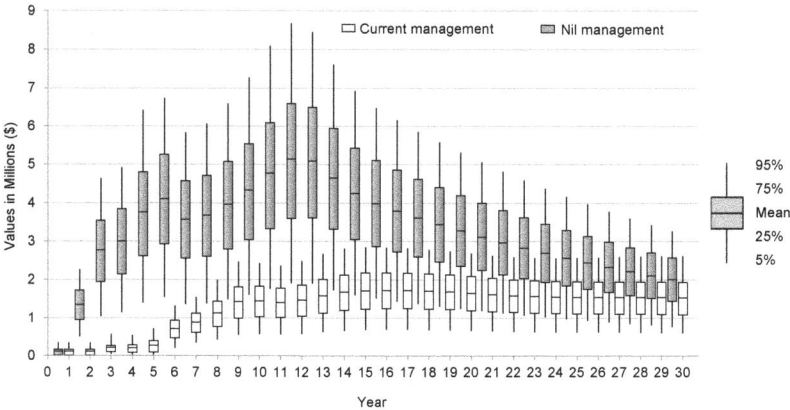

Figure 4.25. Predicted industry losses from rainbow lorikeets in WA under the current and nil management scenarios

---

hollow-nesting fauna through aggressive territorial behaviour (Chapman, 2005; Raidal *et al.*, 1993).

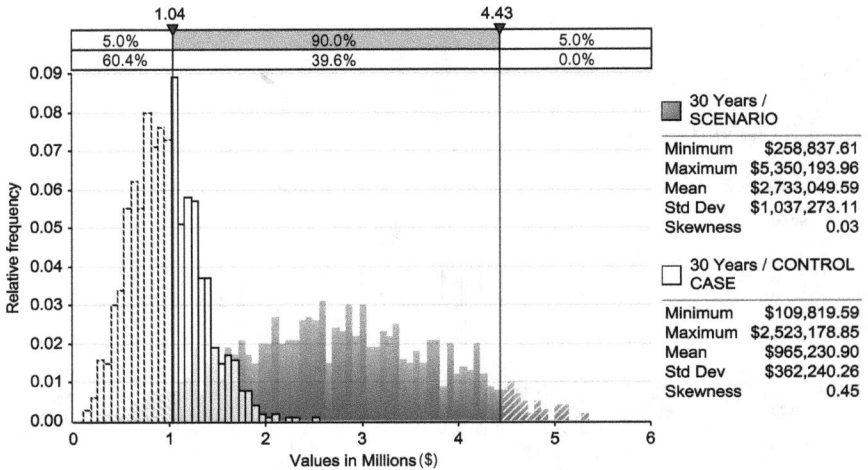

Figure 4.26. Annualised average producer benefit from rainbow lorikeet management in WA

4.4.4.4 *Conclusion*

In this case study we have compared the current rainbow lorikeet management strategy (Rainbow Lorikeet Working Group, 2008) with a nil management scenario in terms of avoided agricultural costs over time. Impacts on pome fruit, stone fruit, citrus, berries, grapes and nuts are considered. If no action is taken to control lorikeets they can be expected to inflict costs to these industries of approximately $2.7 million per annum, but with the management strategy in place this damage can be reduced to $1.0 million. The current management strategy, which is assumed to involve a continued investment of $100,000 over three to five years, is expected to prevent around $1.7 million in damages per annum.

4.4.5 *Summary — vertebrates*

Most vertebrate IAS considered as pests have been present in WA for a long period of time, yet despite being known to producers they still cause large amounts of damage. Because species like the rabbits occupy large areas of the State, these impacts are not localised and

Table 4.9. Rainbow Lorikeet model parameters

Description	Nil management	Current management
Area currently affected, A^{min} (ha). [a]	5.0×10^5	5.0×10^5
Cost of eradication, E (\$/ha). [b]	na	na
Demand elasticity, η. [b]	Uniform(-1.5, -1.3)	Uniform(-1.5, -1.3)
Exponential rate of decline for eradication success probability with respect to area affected, ϕ	na	na
Increased variable cost of production if eradication fails, V (\$/ha). [c]	Discrete([0,106,212],[2.5,0.5,0.25])	Discrete([0,106,212],[1,0.5,0.25])
Intrinsic rate of population density increase, δ (yr^{-1}). [d]	Uniform(0.05,0.10)	Uniform(0.05,0.10)
Intrinsic rate of infestation area growth, r (yr^{-1}). [d]	Pert(0.10,0.15,0.20)	Pert(0.10,0.15,0.20)
Intrinsic rate of satellite generation per unit area of infestation, μ (#/ha). [d]	Pert(5.0×10^{-2}, 7.5×10^{-2}, 1.0×10^{-1})	Pert(5.0×10^{-2}, 7.5×10^{-2}, 1.0×10^{-1})
Maximum area affected, A^{max} (ha). [a]	8.0×10^6	8.0×10^6
Maximum area considered for eradication, A^{erad} (ha)	na	na
Maximum infestation density, K (#/ha). [d]	Uniform(5,10)	Uniform(5,10)
Maximum number of satellite sites generated in a single time step, s^{max} (#). [d]	Pert(10,15,20)	Pert(10,15,20)
Minimum infestation density, N^{min} (#/ha). [d]	Pert(0.00,0.025,0.050)	Pert(0.00,0.025,0.050)

Table 4.9. (*Continued*)

Description	Nil management	Current management
Minimum number of satellite sites generated in a single time step, s^{min} (#).	1.0	1.0
Population diffusion coefficient, D (ha/yr).[d]	Pert(4,6,8)	Pert(2,3,4)
Prevailing price for affected commodities in the first time step, P_0 ($/T).[e]	Berries 6284; Citrus 1554; Nuts 1899; Pome fruit 1193; Stone fruit 2148; Sultanas 3182; Table grapes 3378; Wine grapes 1252	Berries 6284; Citrus 1554; Nuts 1899; Pome fruit 1193; Stone fruit 2148; Sultanas 3182; Table grapes 3378; Wine grapes 1252
Probability of re-entry and establishment, z (%).	na	na
Reinfestation detection probability. Yield reduction despite control, Y (%).[f]	Berries Uniform(0,10); Citrus Uniform(0,10); Nuts Uniform(5,15); Pome fruit Uniform(0,10); Stone fruit Uniform(0,10); Sultanas Uniform(0,10); Table grapes Uniform(0,10); Wine grapes Uniform(0,10)	na; Berries Uniform(0,5); Citrus Uniform(0,5); Nuts Uniform(0,10); Pome fruit Uniform(0,5); Stone fruit Uniform(0,5); Sultanas Uniform(0,5); Table grapes Uniform(0,5); Wine grapes Uniform(0,5)

[a]Based on Chapman (2005) and Rainbow Lorikeet Working Group (2008); [b]Ulubasoglu *et al.* (2011); [c]Assumptions: (i) labour costs of $100/ha (i.e. 3 hr/ha), (ii) ammunition costs equivalent to 10 shotgun cartridges per ha at $0.30 per cartridge (Tracey *et al.*, 2007), (iii) firearm maintenance costs equivalent to $3.00 per ha, and (iv) it is five times more likely that 0 culls will be conducted an affected ha of land by private land managers than 1 cull, and 10 times more likely than 2 culls; [d]Waage *et al.* (2005); [e]ABS (2012). Note "nuts"

are difficult to manage in a coordinated way. We also considered what might happen if an exotic vertebrate species was introduced from a neighbouring State, and the damages potentially caused over time from a native Eastern-Australian species introduced to WA relatively recently. In these case studies we presented results based on two scenarios, one in which minimal biosecurity effort was expended to mitigate the (potential) impact of these species and another in which active management takes place.

If we were to establish priorities on the bases of the first scenario, we would rank wild rabbits as the most costly species (causing $31.9 million worth of damage to plant industries per year), wild dogs as the second most damaging ($13.7 million) and starlings as the lowest of the three ($11.7 million despite not being established in the State at the current time). Priorities would be exactly the same under the active management scenario, although expected damages are marginally lower (i.e. $27.8 million, $13.7 million and $4.8 million, respectively).

4.5 Terrestrial Plants

4.5.1 *Outline*

The precise number of plants introduced to WA is not known, between 2,000 and 3,000 plant species have become established Australia-wide since the arrival of Europeans (Pheloung, 2003; Groves and Hosking, 1998; Groves *et al.*, 2005). Of these, species that compete with pastures and crops for light, water and nutrients present a problem for farmers. Not only do they reduce crop yields but they also contaminate final products, necessitating costly cleaning procedures and sometimes causing price discounts. To reduce their negative impacts, farmers use a range of herbicides which are also costly and expensive to apply. When cropping plant species appear outside farm boundaries, they too are effectively weeds and provide a reservoir for diseases and plant pests that can influence farm productivity.

In this section, we provide three weed case studies involving species that have been introduced to WA. As in Section 4.4, we present two scenarios with the model for each case study; one where biosecurity effort is invested to reduce the damage caused by plant IAS and another where it is not (i.e. nil management). As each of the case studies involves species that have already arrived in WA, existing response costs have been included to provide indicative returns on investment in targeted response efforts.

4.5.2 *Gamba grass (Andropogon gayanus)*

4.5.2.1 *Description*

Gamba grass (*Andropogon gayanus*) is a tall African grass that is drastically altering the understory structure of oligotrophic savannas in tropical Australia (Rossiter-Rachor *et al.*, 2009). It was deliberately introduced into Northern Australia in 1931 by CSIRO to provide cattle fodder, but without proper management has spread rapidly from pastoral plantings in the Northern Territory, Queensland and Western Australia (Reeves *et al.*, 2014).

Gamba grass forms very dense stands up to 4 m high. This compares to native grasses that tend to be more sparsely distributed and grow to approximately 1–3 m. Gamba grass generally has higher photosynthetic rates than native grasses and maintains higher levels of photosynthesis into the dry season (Rossiter *et al.*, 2002). These factors contribute to the grass causing an increased fire risk in areas where it has become abundant.

Gamba grass was formerly recognised as a declared species in WA in 2008 where it is declared C2 Eradication under the *Biosecurity and Agriculture Management Act 2007* (Parliament of Western Australia, 2007). Currently, the weed occupies over approximately 1.5 million ha in the Northern Territory, with the most significant infestations in the Darwin and Katherine regions. In Queensland, it exists over at least 18,000 ha in Cape York Peninsula, with additional populations scattered across coastal and sub-coastal North Queensland.

In WA, gamba grass is restricted to one infestation on El Questro station in the East Kimberley where some 1,770 ha of the weed was planted for cattle fodder in 1991. This infestation has since been under intensive management control. Other infestations have been reported elsewhere in the Kimberly and Pilbara regions and have been eliminated, although they remain under surveillance. Statewide, it is estimated that a total area of 500 ha remains infested with the weed (Reeves *et al.*, 2014).

4.5.2.2 *Parameters*

See Table 4.10.

Table 4.10. Gamba grass model parameters

Description	Nil management	Eradication
Area currently affected, $A^{\min}$ (m^2).[a]	5.0×10^6	5.0×10^6
Cost of eradication, E ($/ha)[a]	0.0	Pert(75, 100, 125)
Demand elasticity, η.[b]	Uniform $(-1.5, -1.3)$	Uniform $(-1.5, -1.3)$
Exponential rate of decline for eradication success probability with respect to area affected, ϕ	Pert$(-0.1, -0.15, -0.2)$	Pert$(-0.1, -0.15, -0.2)$
Increased herbicide and application cost if eradication fails, V ($/ha).[c]	Discrete ([0, 124, 144], [2.5, 0.5, 0.25])	Discrete ([0, 124, 144], [1, 0.5, 0.25])
Intrinsic rate of infestation and density increase, r (yr^{-1}).[d]	Pert(1, 2, 3)	Pert(1, 2, 3)
Intrinsic rate of satellite generation per unit area of infestation, μ (#/m^2).[d]	Pert$(5.0 \times 10^{-2}, 7.5 \times 10^{-2}, 1.0 \times 10^{-1})$	Pert$(5.0 \times 10^{-2}, 7.5 \times 10^{-2}, 1.0 \times 10^{-1})$
Maximum area affected, $A^{\max}$ (m^2).[a]	2.0×10^{11}	2.0×10^{11}
Maximum area considered for eradication, A^{erad} (m^2)	0.0	2.5×10^7
Maximum infestation density, K (#/m^2).[d]	Pert$(1.0 \times 10^4, 5.5 \times 10^4, 1.0 \times 10^5)$	Pert$(1.0 \times 10^4, 5.5 \times 10^4, 1.0 \times 10^5)$
Maximum number of satellite sites generated in a single time step, $s^{\max}$ (#).[d]	Pert(30, 40, 50)	Pert(30, 40, 50)
Minimum infestation density, $N^{\min}$ (#/m^2).	1.0×10^{-4}	1.0×10^{-4}
Minimum number of satellite sites generated in a single time step, $s^{\min}$ (#).	1.0	1.0

(*Continued*)

Table 4.10.   (*Continued*)

Description	Nil management	Eradication
Population diffusion coefficient, D (m^2/yr). [d]	Pert$(0.0, 1.5 \times 10^{-3},$ $2.0 \times 10^{-3})$	Pert$(0.0, 1.5 \times 10^{-3},$ $2.0 \times 10^{-3})$
Prevailing beef price (young steers for export) in the first time step, P_0 ($/kg).[e]	1.75	1.75
Probability of re-entry and establishment, z (%).	1.0	Uniform(20,70)
Reinfestation detection probability.	0.0	Binomial(1.0, 0.6)
Yield reduction despite control, Y ($/ha). [f]	Pert(0.0, 0.4, 0.7)	Pert(0.0, 0.4, 0.7)

[a]Reeves *et al.* (2014). Note 1 ha = 10,000 m^2; [b]Ulubasoglu *et al.* (2011); [c]Assumptions: (i) labour costs of $100/ha (i.e. 1 application × 2 hr/ha × $50/hr); (ii) Glyphosate 450 g/L applied at 10–15 ml/L (i.e. approximately $20/ha); and (iii) it is five times more likely that 0 herbicide treatments will be applied to an affected hectare of land by private land managers than 1 treatment in the absence of an eradication campaign, and 10 times more likely than 2 treatments; and (iv) fire and emergency service provision equivalent to approximately $4/ha (based on estimates from the Shire of Coomlie Shire (165,000 ha) in the Northern Territory (Reeves *et al.*, 2014, Setterfield *et al.*, 2009); [d]Waage *et al.* (2005). Note 1 ha = 10,000 m^2; [e]Curtis and McCormick (2012); [f]Assumes a yield loss of 0–2% in both the nil management and eradication scenarios and a beef gross margin of approximately $35/ha (McCosker *et al.*, 2010).

4.5.2.3 *Results*

See Figures 4.27–4.29.

4.5.2.4 *Conclusion*

The impact of gamba grass on WA agriculture is sufficiently large to warrant an eradication investment of up to $16.8 million per year. This is indicative of a very serious threat to the State, a fact largely attributable to the large potential area affected by the weed and increased fire risk for these areas if it becomes widely established. If no biosecurity effort is invested in its control, gamba grass is predicted to cause damages of approximately $110.0 million per year over the next 30 years. This impact can be reduced to $93.2 million with the continuation of current management activities.

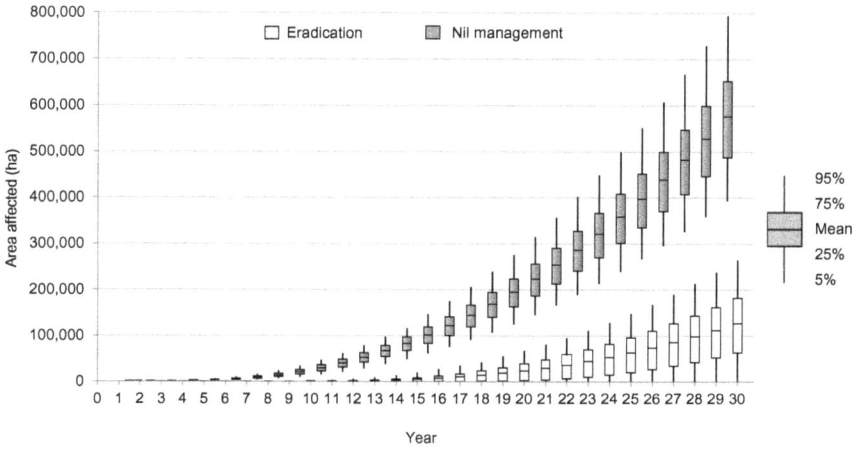

Figure 4.27. Expected area affected by gamba grass in WA under eradication and nil management scenarios

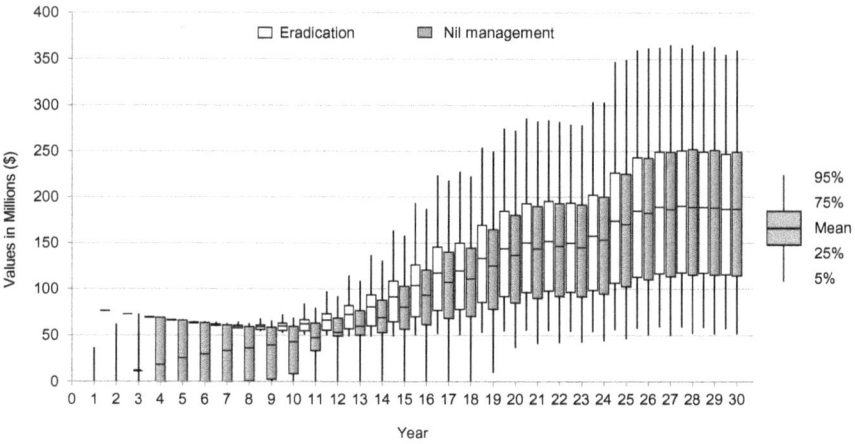

Figure 4.28. Predicted industry losses from gamba grass in WA under the eradication and nil management scenarios

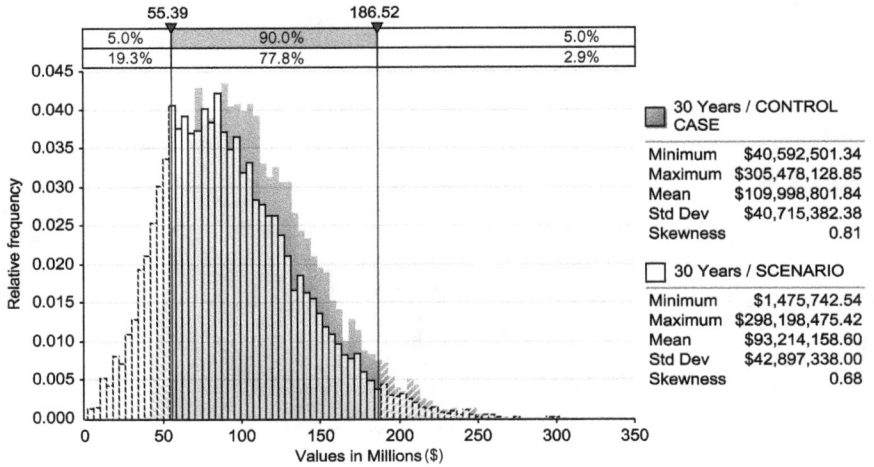

Figure 4.29. Annualised average producer cost attributable to gamba grass under an eradication and a nil management scenario

4.5.3 *Rubber vine (Cryptostegia grandiflora)*

4.5.3.1 *Description*

Rubber vine (*Cryptostegia grandiflora*) is a highly invasive woody perennial vine that displaces native vegetation by preventing access to adequate sunlight. The plant forms impenetrable thickets with dense canopies preventing both human and animal access to affected areas. Rubber vine is native to Madagascar and was introduced to Australia as an ornamental plant in the 1860s (Parsons and Cuthbertson, 1992). Widespread planting occurred in the 1940s when the plant was considered to have commercial potential as a source of rubber, hence the name rubber vine (Tomley and Evans, 2004). As a result, the plant was also introduced to other tropical areas around the world (Tomley, 1995).

In Australia, the vast majority of infestation occurs in Queensland where it is estimated to infest over 700,000 ha (Australian Weeds Committee, 2012). Its spread continues, with most of tropical and sub-tropical Northern Australia comprising of suitable rubber vine habitat (Australian Weeds Committee, 2012). In WA, rubber vine is only known to occur on the lower Fitzroy River and lower Lake Argyle Catchment in the Kimberley region. It is assumed that the remainder of the Kimberly and most of the Pilbara region could potentially be affected (CRC for Australian Weed Management, 2003).

The main agricultural impacts of rubber vine are associated with reduced grazing capacity and control costs. The plant is only eaten by cattle when dense growth surrounds water courses, when fresh regrowth is present following burning, or when other feed is unavailable (Cook *et al.*, 1990). It is toxic to livestock and stock losses have been known to occur as a result of excessive grazing, with horses being particularly susceptible (McGavin, 1969). However, in this assessment we only deal with producer losses attributable to pasture loss and control costs.

It is believed around 6,000 ha of the Kimberley region was affected by rubber vine before concerted response efforts were mounted by community groups in 2012. At this time, Rangeland NRM received a $1.0 million grant to control the weed over a 3-year period in partnership with the Lake Argyle Rubber Vine Advisory Committee, the West Kimberley Rubber Vine Steering Committee and the Kimberley Rangelands Biosecurity Association infestation (Rangelands NRM, 2013). DAFWA are also involved in an advisory capacity. Upon completion the project aims to have achieved a 90% reduction in the WA rubber vine infestation (Rangelands NRM, 2013). This assessment assumes it has been successful and that the current infestation (as the project nears completion) is between 500 ha and 700 ha, and that eradication or remaining plants is a viable option for future control.[4]

---

[4]Two biological agents have had an impact in reducing the health and spread of rubber vine in Australia. The pyralid moth (*Euclasta whalleyi*) was released in Queensland between 1988 and 1991 but populations have remained small due to pressure from generalist parasitoids (Mo *et al.*, 2000). Rubber vine rust (*Maravalia cryptostegiae*) was then released in 1995 and met with some success (Tomley and Evans, 2004). However, neither can be relied upon to kill mature rubber vine plants on their own (CRC for Australian Weed Management, 2003), and are not considered in this assessment.

4.5.3.2 *Parameters*

See Table 4.11.

Table 4.11. Rubber vine model parameters

Description	Nil management	Eradication
Area currently affected, $A^{\min}$ (m^2).[a]	Uniform(5.0 × 10^6, 7.0 × 10^6)	Uniform (5.0 × 10^6, 7.0 × 10^6)
Cost of eradication, E ($/ha).[a]	0.0	1.5 × 10^5
Demand elasticity, η.[b]	Uniform (−3.2,−2.0)	Uniform (−3.2,−2.0)
Exponential rate of decline for eradication success probability with respect to area affected, ϕ	Pert(−0.1, −0.15,−0.2)	Pert(−0.1, −0.15,−0.2)
Increased herbicide and application cost if eradication fails, V ($/ha).[c]	0 to 65	0 to 65
Intrinsic rate of infestation and density increase, r (yr^{-1}).[d]	Pert (0.50,0.75,1.00)	Pert (0.50,0.75,1.00)
Intrinsic rate of satellite generation per unit area of infestation, μ (#/m^2).[d]	Pert(1.0 × 10^{-5}, 5.95 × 10^{-4}, 1.0 × 10^{-3})	Pert(1.0 × 10^{-5}, 5.95 × 10^{-4}, 1.0 × 10^{-3})
Maximum area affected, $A^{\max}$ (m^2).	Kimberley 1.4 × 10^{11} Pilbara 1.7 × 10^{11}	Kimberley 1.4 × 10^{11} Pilbara 1.7 × 10^{11}
Maximum area considered for eradication, A^{erad} (m^2)	0.0	Pert(1.5 × 10^7, 2.0 × 10^7, 2.5 × 10^7)
Maximum infestation density, K (#/m^2).[d]	Pert(1.0 × 10^4, 5.5 × 10^4, 1.0 × 10^5)	Pert(1.0 × 10^4, 5.5 × 10^4, 1.0 × 10^5)
Maximum number of satellite sites generated in a single time step, $s^{\max}$(#).[d]	Uniform(10,20)	Uniform(10,20)
Minimum infestation density, $N^{\min}$ (#/m^2).	1.0 × 10^{-4}	1.0 × 10^{-4}
Minimum number of satellite sites generated in a single time step, $s^{\min}$ (#).	1.0	1.0
Population diffusion coefficient, D (m^2/yr).[d]	Pert(5.0 × 10^3, 6.25 × 10^3, 7.5 × 10^3)	Pert(5.0 × 10^3, 6.25 × 10^3, 7.5 × 10^3)

(Continued)

Table 4.11. (*Continued*)

Description	Nil management	Eradication
Prevailing price for affected commodities in the first time step, P_0 ($/T).[e]	1.75	1.75
Probability of re-entry and establishment, z (%).	100	Uniform(21,70)
Reinfestation detection probability (%).	0.0	Binomial(100,0.4)
Reinfestation detection probability >1 yr after successful eradication (%).	0.0	Binomial(100,0.8)
Yield reduction despite control, Y (%).[f]	Uniform(0.0,0.2)	Uniform(0.0,0.2)

[a]Based on Rangelands NRM (2013). Note 1 ha $= 10,000\,m^2$. Also note in the case of eradication costs that the initial knockdown phase (i.e. eradicating the current 500–700 ha infested) is arbitrarily assumed to take between one and five years to achieve (i.e. discrete([1,2,3,4,5],[0.1,0.5,1.0,1.0,0.5])); [b]Ulubasoglu *et al.* (2011); [c]Assumptions: (i) labour costs of $100/ha (i.e. one application × 2 hr/ha × $50/hr); (ii) chemical costs equivalent to triclopyr 240 g/L plus picloram 120 g/L at a rate of 1 L/60 L diesel mix (Doak *et al.*, 2004) (i.e. $200/ha); and (iii) it is two times more likely that zero herbicide treatments will be applied to an affected hectare of land by private land managers than one treatment in the absence of an eradication campaign, and four times more likely than two treatments; [d]Specified with reference Waage *et al.* (2005); [e]Price is specified using the price of young steers for export in Curtis and McCormick (2012); [f]Assumes a yield loss of 0–2% in both scenarios and a beef gross margin of approximately $35/ha (McCosker *et al.*, 2010).

4.5.3.3 *Results*

See Figures 4.30–4.32.

4.5.3.4 *Conclusion*

Rubber vine poses a very large risk to pastoral production in the tropical and sub-tropical north of the State. The weed's relatively high rate of growth and huge potential area of infestation mean that the costs associated with no management can be very high indeed. It is estimated that the eradication of the current infestation would prevent damages of $8.7 million per year. This is based on agricultural impacts of the weed, and ignores potentially large ecological costs. If eradication and maintenance of area freedom can

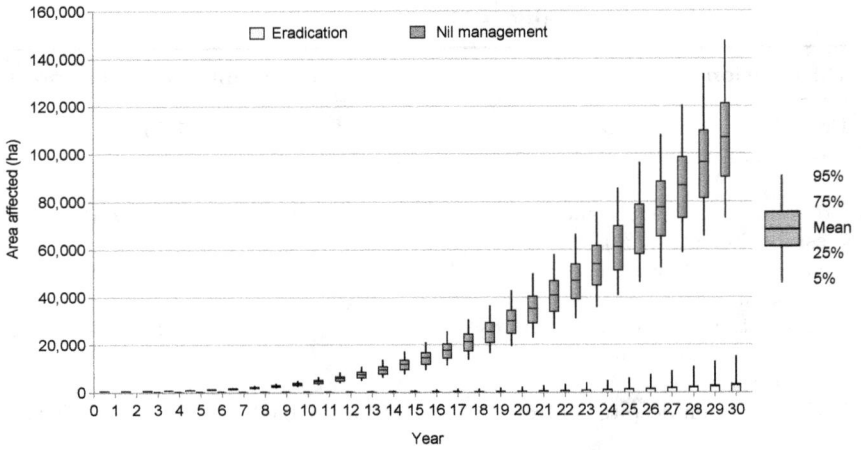

Figure 4.30. Expected area affected by rubber vine in WA under eradication and nil management scenarios

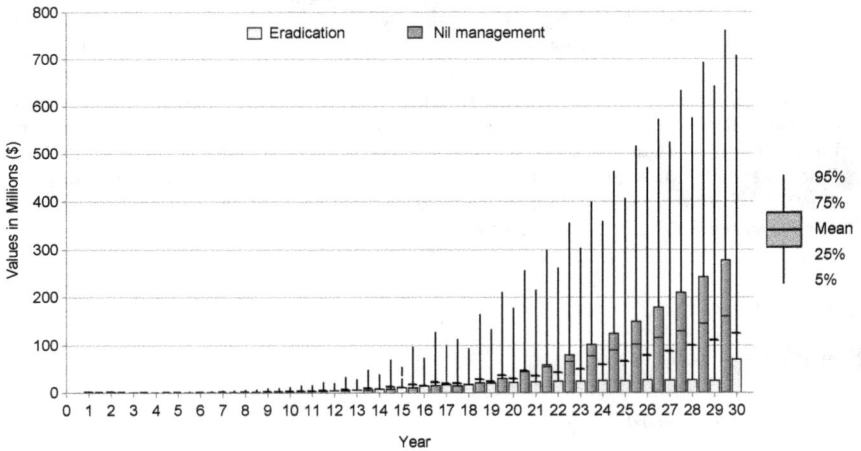

Figure 4.31. Predicted industry losses from rubber vine in WA under the eradication and nil management scenarios

be maintained for less than this amount, substantial gains will have been achieved for the agricultural economy.

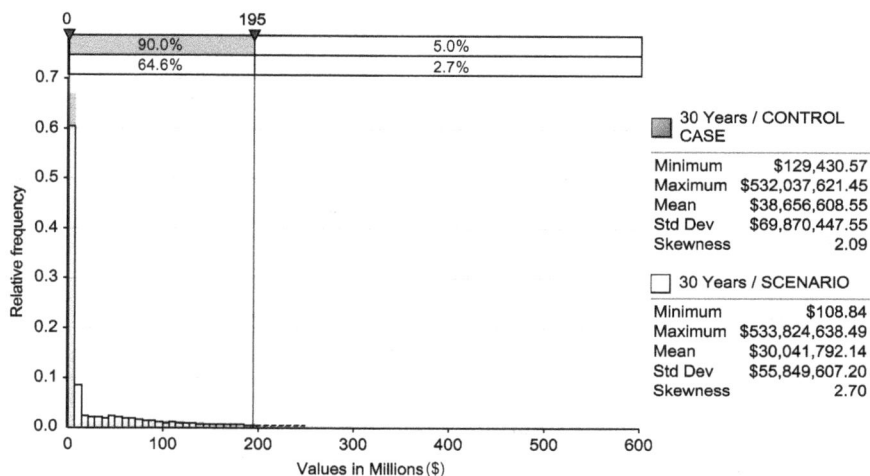

Figure 4.32. Annualised average producer cost from rubber vine in WA over a 30-year period under a nil management scenario and an eradication scenario

4.5.4 *Three-horned bedstraw* (*Galium tricornutum* Dandy)

4.5.4.1 *Description*

Three-horned bedstraw (*Galium tricornutum* Dandy), henceforth termed bedstraw, is a weed of broad acre crops that is established in the Southeast of Australia. Until recently it was considered absent from WA. A native of eastern or southern Europe, it is a close relative of another well-known weed species, cleavers (*G. aparine*) (Moore and Dodd, 2008). Bedstraw is a competitive climbing plant that forms dense masses of tangled vegetation in crops, along fence lines and in other disturbed agricultural land (Edward and Kingwell, 2003; DAWA, 1997). It displaces crops and contaminates harvests, particularly in the case of canola due to the similarity of the seeds of both plants making it difficult to remove by screening.

Bedstraw has been detected on several occasions in WA, beginning in the Shire of Boddington in 1968 where three bedstraw plants were collected. It was not detected again until it was found on the Department of Agriculture and Food research station in the Shire of Mount Barker in 2001 where it was eradicated. The most recent detections have occurred in the shires of Darkan (2003), Merredin (2004) and Cranbrook (2005), all of which are considered to be independent populations that originated from separate incursions (DAWA, 1997).

Following the Merredin detection, an economic analysis of a possible eradication campaign was undertaken (Edward and Kingwell (2003)) that concluded a net benefit of almost $600 million could be achieved over a period of 12 years if eradication was successful.[5] An eradication program was subsequently commenced in late 2003 and still remains in place. It is now funded by the *Grains, Seed & Hay Industry Funding Scheme* established under the *Biosecurity and Agriculture Management Act 2007* (Parliament of Western Australia, 2007), which has made approximately $150,000 available for the eradication program in 2013/2014. Currently, around 420 ha are infested with bedstraw.

4.5.4.2 *Parameters*

See Table 4.12.

4.5.4.3 *Results*

See Figures 4.33–4.35.

4.5.4.4 *Conclusion*

Bedstraw presents a potential threat to WA broad acre industries, and due to their large production area the effects of crop displacement can be expected to accumulate over time if no management action had been taken. The impact of the weed on wheat, barley, oats, lupin and canola producers, particularly in the long term, is sufficiently large to warrant an eradication investment of up to $1.3 million per year.

---

[5]The eradication benefits estimated by Edward and Kingwell (2003) are substantially higher than those put forward here. This is primarily due to two assumptions used in their analysis: (i) exponential bedstraw growth of 40% per annum, and; (ii) high yield loss (i.e. 5% in acid soils and 35% in alkaline soils). Despite both spread and yield losses being assumed smaller in our case study analysis, the benefits of eradication (revealed below) remain high.

Table 4.12. Three-horned bedstraw model parameters

Description	Nil management	Eradication
Area currently affected, A^{min} (m²). [a]	4.2×10^6	4.2×10^6
Cost of eradication, E ($/ha). [a]	0.0	1.5×10^5
Demand elasticity, η. [b]	∞	∞
Exponential rate of decline for eradication success probability with respect to area affected, ϕ	Pert($-0.1, -0.15, -0.2$)	Pert($-0.1, -0.15, -0.2$)
Increased herbicide and application cost if eradication fails, V ($/ha). [c]	0 to 30	0 to 30
Intrinsic rate of infestation and density increase, r (yr^{-1}). [d]	Pert(1,2,3)	Pert(1,2,3)
Intrinsic rate of satellite generation per unit area of infestation, μ (#/m²). [d]	Pert(1.0×10^{-5}, 5.95×10^{-4}, 1.0×10^{-3})	Pert(1.0×10^{-5}, 5.95×10^{-4}, 1.0×10^{-3})
Maximum area affected, A^{max} (m²). [e]	7.6×10^{10}	7.6×10^{10}
Maximum area considered for eradication, A^{erad} (m²)	0.0	Pert(5.0×10^6, 7.5×10^6, 1.0×10^7)
Maximum infestation density, K (#/m²). [d]	Pert(1.0×10^4, 5.5×10^4, 1.0×10^5)	Pert(1.0×10^4, 5.5×10^4, 1.0×10^5)
Maximum number of satellite sites generated in a single time step, s^{max} (#). [d]	Uniform(5,10)	Uniform(5,10)

(Continued)

Table 4.12. (*Continued*)

Description	Nil management	Eradication
Minimum infestation density, N^{min} (#/m²).	1.0×10^{-4}	1.0×10^{-4}
Minimum number of satellite sites generated in a single time step, s^{min} (#).	1.0	1.0
Population diffusion coefficient, D (m²/yr).[d]	Pert(5.0×10^3, 6.25×10^3, 7.5×10^3)	Pert(5.0×10^3, 6.25×10^3, 7.5×10^3)
Prevailing price for affected commodities in the first time step, P_0 ($/T).[e]	Barley 215 Canola 540 Oats 175 Lupins 260 Wheat 280	Barley 215 Canola 540 Oats 175 Lupins 260 Wheat 280
Probability of re-entry and establishment, z (%).	100	Uniform(2,21)
Reinfestation detection probability (%).	0.0	Binomial(100,0.4)
Reinfestation detection probability >1yr after successful eradication (%).	0.0	Binomial(100,0.8)
Yield reduction despite control, Y (%).[f]	Uniform(0.0,0.5)	Uniform(0.0,0.5)

[a]Based on Moore and Dodd (2008). Note 1 ha = 10,000m². Also note in the case of eradication costs that the initial knockdown phase (i.e. eradicating the current 420 ha infested) is arbitrarily assumed to take between one and five years to achieve (i.e. discrete([1, 2, 3, 4, 5],[0.1, 0.5, 1.0, 1.0, 0.5])); [b]Host products are predominantly sold to export market and WA is not sufficiently large to exert pressure on the world price; [c]In wheat, barley and oats 1,000 mL/ha application of Starane before the bolt stage at a cost of approximately $16/ha. In lupins, seed cleaning at a cost of $10.15/ha (i.e. $15/T) required for one consignment in 10 (i.e. binomial(1.0,0.1)). In canola, a 1% price deduction applies for each 1% of impurities up to 4%, and a 2% deduction for each 1% of impurities over 4% (AOF, 2012) up to a maximum of 6%; [d]Specified with reference Waage *et al.* (2005); [e]ABS (2012).

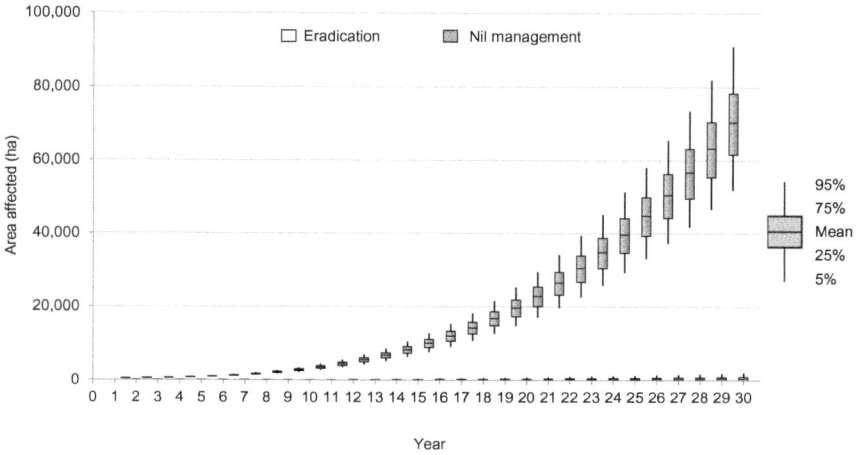

Figure 4.33. Expected area affected by three-horned bedstraw in WA under eradication and nil management scenarios

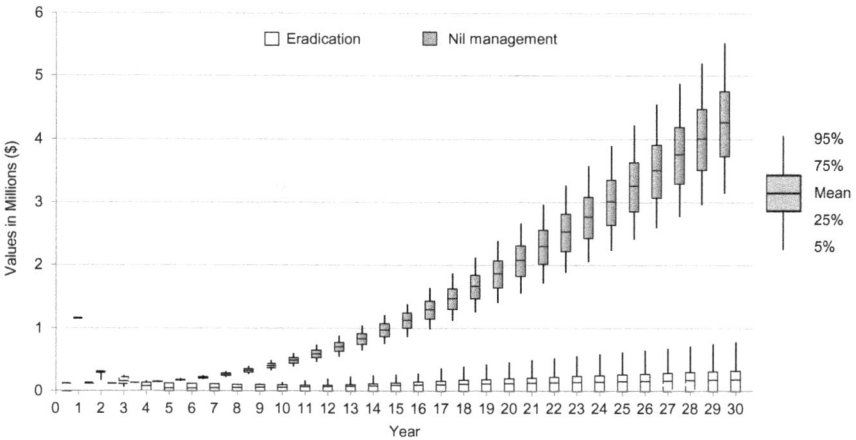

Figure 4.34. Predicted industry losses from three-horned bedstraw in WA under the eradication and nil management scenarios

Plant Biosecurity Policy Evaluation

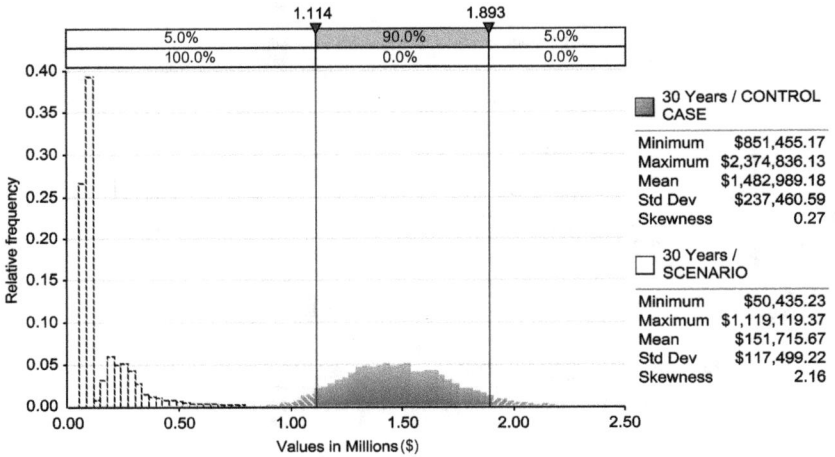

Figure 4.35. Annualised average producer cost from three-horned bedstraw in WA over a 30-year period under a nil management scenario and an eradication scenario

4.5.5 *Summary — terrestrial plants*

We have seen in this section that terrestrial plant introductions have the potential to cause severe damages to WA agriculture. Of the case studies looked at, we would prioritise gamba grass as the highest biosecurity priority on the basis of damages caused over time under the nil management scenario ($110.0 million per year), followed by rubber vine ($38.7 million) and three-horned bedstraw ($1.5 million). Priorities would be the same under the current management scenario, but damages are expected to be reduced to $93.2 million, $30.0 million and $0.2 million, respectively.

4.6 Conclusion

In this chapter we have demonstrated how the bioeconomic model of Chapter 3 can be applied to a range of IAS from different taxanomic groups to estimate their likely impacts to a region over time. The set of IAS we have looked at were assessed in terms of their potential damage cost to agricultural industries in the State of WA over a period of 30 years. Although it is difficult to generalise or to draw definite findings from this relatively small group of IAS, in the next chapter we will try to do so and put forward some speculative ideas about the types of IAS impacts we can expect to observe.

In terms of the species case studies we have examined, covering both established and exotic IAS ranging from invertebrates, plant pathogens, vertebrates and plants, we have been able to predict how the impact each is expected to have on WA agriculture change over time. If we were to prioritise this list of species on the basis of their potential cost to agriculture over 30 years, we would suggest gamba grass, rubber vine and Ug99 have the largest capacity to impose costs, and therefore warrant attention from policymakers. In the case of Ug99, this is a species that remains exotic to the region. Our ranking might change again if we consider priorities on the basis of impacts should IAS be present in the first time period of the analysis. If this were the case, plant pathogens with export implications for major export commodities will be ranked highest, but we must keep in mind that the probability of this event can be remote.

Even when we do take the probability of an IAS arriving into account in our calculations, it is interesting to note how priorities might change if we alter the length of time we are concerned about when communicating risks to policymakers. For instance, if it is a much smaller timeframe we are interested in, say 10 years, our top three species in order of their potential impacts become Ug99, gamba grass and wild rabbits. Over 20 years this would change again to gamba grass, Ug99 and then rubber vine. It follows that the timeframe is critical in determining priorities, yet it is not clear what an appropriate timeframe should be when conducting these sorts of assessments.

The case studies included in this section have only considered economic damage to agriculture, or other monetised damage to industries reliant on plants for one purpose or another. So, environmental and social impacts are not considered; and in fact the modelling approach we have used here to assign financial values to impacts over time is inappropriate for these kinds of non-monetised impacts. In Chapter 7 we will show how a different set of techniques can be used to prioritise IAS management when not just economic but social and environmental impacts are taken into account.

CHAPTER 5

VARYING PATTERNS OF IMPACT OVER TIME: THE CHOICE OF TIME HORIZON

5.1 Introduction

The case studies presented in Chapter 4 are a tiny sample of the invasive alien species (IAS) threatening agricultural industries, the environment and societies around the world. As we have seen through, there is great diversity in how they cause impacts over time. So, what can such a small sample of IAS reveal about all the other species we have not looked at? Well, that is a fair question. The idiosyncrasies of each IAS really mean policymakers should view each on a case-by-case basis, rather than generalise.

However, our tendency towards scientific reductionism compels us to at least attempt to form patterns of impacts we might expect from different groups of IAS. Using the specific case studies, this chapter examines the cost–time relationships to give a rough idea of what costs are imposed by IAS, and when they are imposed. By calculating the total costs of different IAS we can make financial comparisons between species and prioritise them according to their capacity to cause damage over specific time horizons.

If we plot the total cost–time relationship for different IAS it soon becomes apparent that the length of the timeframe we are considering is critical in determining priorities. This chapter describes the types of total cost–time relationships we tend to encounter. As we will show, costs don't necessarily depend on how an IAS spreads, or even the yield losses they cause. Rather, the types of impacts of IAS and the process of discounting future values are the dominant factors determining how costs accrue over time.

5.2 Comparing Patterns of Impact Over Time

5.2.1 *Constant marginal cost over time*

For the majority of IAS affecting plant-based industries, the relationship between the value of costs caused and time is more or less linear, implying that the intertemporal marginal costs of IAS are constant. That is, the degree to which total costs change between time periods remains much the same. Figure 5.1 provides an abstract view of the flow of IAS costs over time (C) for these types of IAS where the change in total cost is identical between years 0–10 (ΔC_{10}), 10–20 (ΔC_{20}) and 20–30 (ΔC_{30}). The resultant total cost curve is linear (C_A).

In general, IAS demonstrating cost curves of this nature tend to be associated with crops of one form or another. Indeed, if we leave aside the 'incursion year 1' scenarios used for exotic IAS in Sections 4.2 and 4.3, all of the cost curves examined in Chapter 4 are linear, or at least approaching linearity. If we ignore the probability of entry

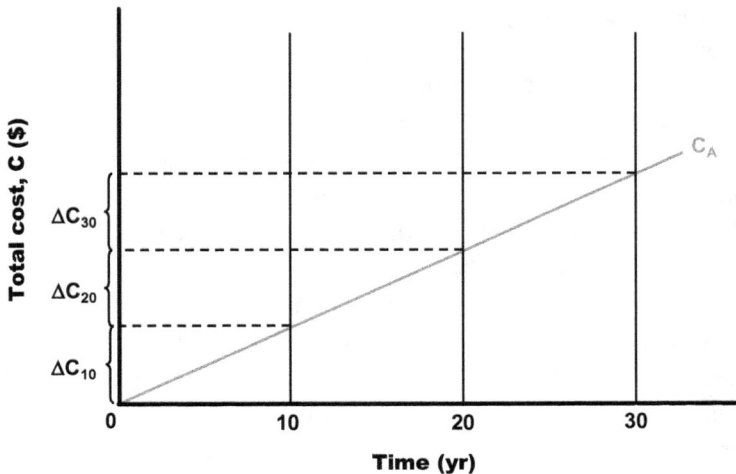

Figure 5.1.   Linear cost–time

and establishment, which is always an important determinant of C, the parameters that determine the costs of these IAS tend to be biological in nature. In particular, the intrinsic rate of spread (r) and diffusion coefficients (D) are highly influential in shaping the total cost curve for these species (Waage *et al.*, 2005; Waage and Mumford, 2008). As we shall see, this is not always the case.

It should be pointed out that a linear C curve does not imply a linear biological spread pattern. The reason why we observe C curves that are roughly linear is largely attributable to the process of discounting future impacts. We discussed discounting briefly in Chapter 3 when introducing the economic model, and we revisit it here to highlight discounting's significance when making species comparisons.

In short, higher discount rates mean future IAS impacts are discounted more heavily and are of less importance to any decision maker concerned with market-based impacts. We will return to balance this issue with the consideration of non-market impacts in Chapter 7. For now, let us simply assume both *economic* and *social/environmental* impacts are captured by the C curve of Figure 5.1. Both are discounted, but for different purposes.

A discount rate is applied to economic impact calculations to reflect the opportunity cost of investment decisions, but the *social* discount rate is harder to define. In the social or government context, discounting reflects the view that future generations will be better off than the current generation. Technological progress is making the production of goods and services cheaper over time, and real incomes are rising. So, \$100 is worth less to a person living in the society of 20 years' time than it is to the average person today. Accordingly, the social benefits of biosecurity policies accruing in the future (or the value of prevented IAS costs) should be discounted (Waage *et al.*, 2005). The question is, how much less will a person of the future value today's money, and therefore what is the appropriate social discount rate to apply to our calculations?

In the absence of clear information on opportunity costs relevant to specific projects (e.g. like the control IAS), economists tend to cite government discount rate 'guidelines'. In Australia, there is no prescribed discount rate to use in the analysis of biosecurity projects *per se*, but for publicly funded projects it consists of a margin on top of a private discount rate of around 3% (Commonwealth of Australia, 2006). This margin, which we arbitrarily assume is 2%, reflects the costs incurred by society in the transfer of money from the private sector to the public sector (i.e. via taxes). Hence, in the case studies of Chapter 4 we used a constant discount rate of 5%. Of course, not all biosecurity investments are made by government, so the choice of discount rate, like so many things in biosecurity, is uncertain.

5.2.2 *Diminishing marginal cost over time*

Pests and diseases whose damages are less gradual have distinctly different cost–time relationships associated with them. In some cases, the costs of IAS are concentrated early-on in the invasion process and gradually taper off over time. Figure 5.2 provides an abstract view of costs over time for these types of IAS. Here, the change in total costs between years 0 and 10 (ΔC_{10}) is larger than in the 10–20-year interval (ΔC_{20}), which in turn is larger than the change in costs over the 20–30-year interval (ΔC_{30}). The resultant total cost curve is labelled (C_B).

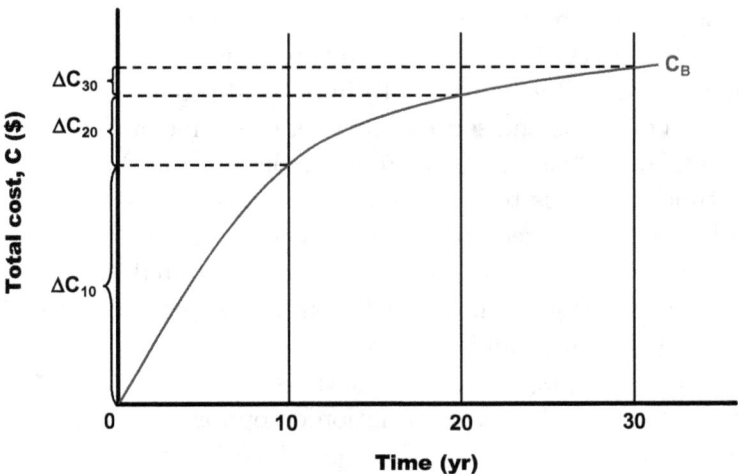

Figure 5.2. Concave cost–time

Model parameters of the most significance in these kinds of cost–time relationships tend to be the probability of entry and establishment and the reduction in export earnings attributable to a loss of pest/disease area freedom (Waage *et al.*, 2005; Waage and Mumford, 2008). If the costs of export market losses are felt immediately upon detection, the total cost on the economy between the years 0 and 10 (depending on the assumed probabilities of entry and establishment) are severe. This explains the rapid growth in costs during this time interval. Thereafter, costs accrue at a slower and slower rate, resulting in the concave cost curve of C_B in Figure 5.2.

Of the case studies we looked at in Chapter 4, the clearest examples of cost curves approaching a concave shape can be found in 'incursion year 1' scenario C curves for the broadacre crop invertebrate pests (cabbage seedpod weevil, khapra beetle and wheat stem sawfly) and pathogens (Ug99, barley stripe rust and Karnal bunt). In each of these cases, costs are very high immediately upon entry despite IAS abundance being low (see Sections 4.2 and 4.3).

5.2.3 *Increasing marginal cost over time*

A third type of curve is encountered that exhibits increasing intertemporal marginal costs. This curve is typical of 'sleeper' species (Randall, 2002); IAS that take long periods to establish and spread beyond a low level, but when they eventually do they can cause a huge amount of damage. Of the examples shown in Chapter 4, gamba grass and rubber vine fall into this category (see Section 4.5).

Sleeper species often have environmental impacts. For instance, sleeper weeds can be present within the environment and remain relatively benign for long periods of time before becoming a problem, but by that time it is difficult for any management action to be taken. This can be very damaging in terms of biodiversity, particularly when these species out-compete native vegetation for nutrients and sunlight and eventually form monocultures. If the value of non-market costs could be included in our assessment of costs over time, they would further exacerbate this type of impact-time curve, but this is seldom done in quantitative studies of IAS.

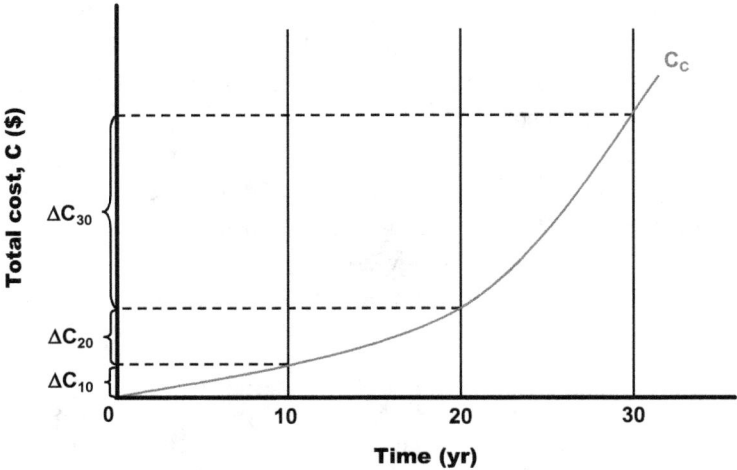

Figure 5.3. Convex cost–time

Figure 5.3 demonstrates the impact of these effects conceptually. Consider the case of an IAS with a negative effect on environmental goods. The change in the expected consequence of such species between years 0 and 10 (ΔC_{10}) is less that that occurring between years 10 and 20 (ΔC_{20}), which in turn is less than that occurring between years 20 and 30 (ΔC_{30}). The slope of the resultant total cost curve (C_C) is positive and increasing at an increasing rate. Note that this convex curve shape implies that the effect of the income elasticity of demand and price/value increases for environmental goods overrides the effects of discounting.

5.3 Cross-over Effects and Uncertainty

Having described three cost–time relationships, it is possible to speculate how a policymaker might begin to make decisions and compare between IAS according to the accumulation of total damage costs over time. Conceivably, cost-minimising policymakers might be faced with situations where desirable strategies are quite different for different time horizons. In particular, the total cost curves related to different species could cross over one another at some point in time,

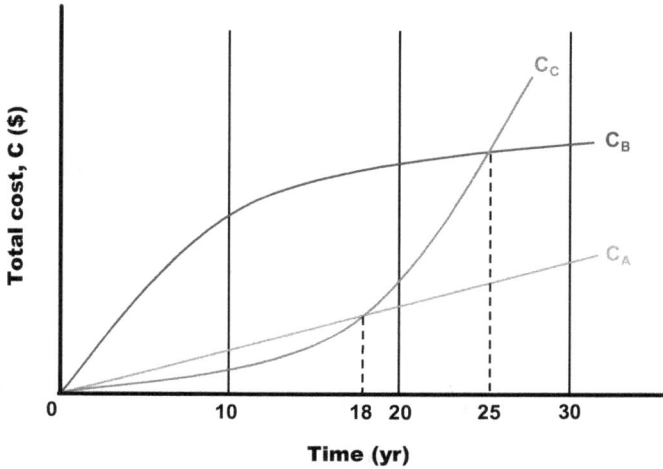

Figure 5.4. Cross-over effects

making it critical to specify what period of time is being considered at the outset of the analysis (Cook *et al.*, 2011b).

Assume, for instance, a policymaker is concerned with three hypothetical IAS, A, B and C, characterised by the respective expected cost curves C_A, C_B and C_C shown in Figure 5.4. Further, assume they have sufficient information to track the flow of total costs accurately over time, including non-market impacts. So, variance around the mean is negligible. The choice of risk management strategy will largely depend on the length of time policymakers consider relevant.

If an appropriate time horizon for biosecurity policymaking were 10 years, a consequence-minimising strategy would involve the targeting of IAS B. The explosive impact of this IAS early in the time horizon gives it high biosecurity significance in the short term. By making its exclusion a policy priority, the invasive cost is minimised. Pests A and B are less significant in the short term, so there appears to be little strategic merit in targeting biosecurity policies towards them over a 10-year timeframe. By year 18, the C_C curve crosses the C_A curve as the presence of IAS C begins to cause substantial damage, but by year 20 the benefits of targeting

IAS B still outweigh those of other IAS. However, by years 25–30 the situation is quite different. The C_C curve has now crossed over the C_B curve, intersecting it from below in year 25. Hence, IAS C now represents the species of greatest biosecurity significance to the region. As a consequence, the time horizon for policymaking clearly determines the priority for resource allocation.

If species A and B can be thought of as typical diseases of crops, while species C represents an environmental invasive, then the optimal risk management strategy begins to look familiar. In the short term, the explosiveness of fungal pathogens creates a great deal of policy interest. However, over a longer time horizon, the significance of other IAS can increase dramatically. A failure to control such IAS early effectively passes the burden of control on to future generations. This explains why many environmental IAS are overlooked by biosecurity policies for considerable periods of time (Simberloff, 2006). Once they begin to feature prominently in biosecurity risk profiling for specific regions, as is the case for IAS C by the 25-year mark of Figure 5.5, they can be extremely difficult to control and impossible to eradicate.

Moreover, predictive modelling of IAS introduced to new regions involves the use of broadly defined parameter distributions rather than point estimates. This means that there are a range of possible policy considerations that might or might not occur. In the case of our hypothetical species A, B and C, this uncertainty makes the decision of which to treat as a biosecurity priority more complicated, as Figure 5.5 shows.

If only the average cost–time curves are considered the situation has not changed from Figure 5.4 as far as the decision maker is concerned. IAS B has the highest invasion impact until year 25 after which it is surpassed by IAS C, and IAS A has the second highest impact up until year 18 when it is surpassed by IAS C. However, if the variability of impact estimates is taken into account, as indicated by the broken lines either side of C_A, C_B and C_C then prioritising becomes dependent on the decision maker's attitude to biosecurity risk and uncertainty (Waage *et al.*, 2005).

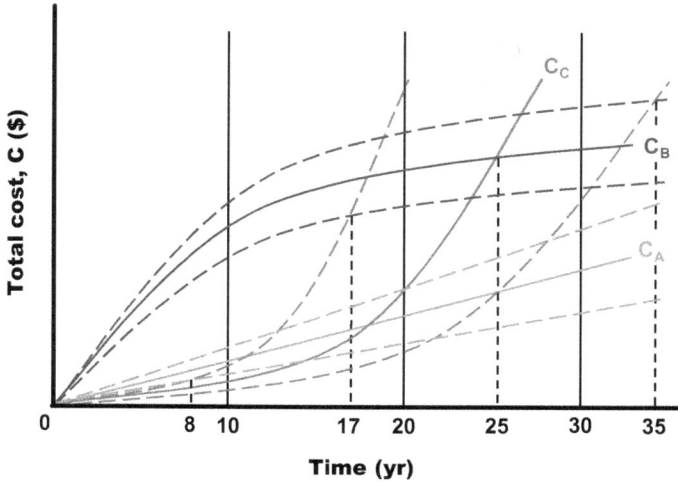

Figure 5.5. Uncertainty and consequence assessment

For instance, the impact of IAS C may exceed that of B by as early as year 17, so a decision made over a 20-year time horizon by a risk-averse policymaker may lead to C being prioritised for biosecurity resources over B. Similarly, the impact of A may be exceeded by that of IAS C as early as year 8. On the other hand, the cross-over between IAS C and B may occur as late as year 35. In light of this information, a policymaker looking over a 20-year time horizon may choose to prioritise resources towards the management of B at the expense of A and C.

5.4 Conclusion

When predicting the economic, social and environmental costs of IAS as they impact in new situations we face scientific uncertainties in the process of spread and political/regulatory uncertainties concerning the length of time we are interested in. These complications mean that prioritising resources to high-cost species is difficult. This is particularly true when the total costs of an IAS include environmental and social impacts as well as economic impacts.

To explicitly include social and environmental IAS impacts into biosecurity policy analysis may require us to look beyond conventional price-based assessment methods. Rather than concentrating on valuing such (non-market) impacts so that they are comparable to economic (market) impacts using financial measures, it may be a case of converting monetary estimates of economic impacts into semi-quantitative or qualitative terms so that they can be compared with social and environmental impacts. This is a somewhat controversial idea, but one that deserves our attention for pragmatic reasons, not just philosophical ones. We will discuss this in full in Chapter 7.

Before doing so, however, we must deal with another issue embedded within our assessment model. This issue is of particular relevance to exotic IAS, or IAS that have not previously been established within a region. It concerns the nature, scale and success of responses to IAS, which impacts on the total cost curves discussed in this chapter. After all, if an IAS is relatively easy to remove from a region once it is detected, we expect this to be reflected in a lower cost–time curve. On the other hand, if it is difficult to detect and/or remove, the opposite may be true (unless the reason it is difficult to detect is because its cost and revenue implications are relatively minor), and the best strategy may be to learn to live with it once prevention has failed. This issue about prevention vs. control is the focus of the next chapter.

CHAPTER 6

PREVENTION VS. ERADICATION

6.1 Introduction

In this chapter, we contrast eradication and prevention strategies. Studies of IAS often assert that "prevention is better than cure" on the basis that it is the least cost risk-mitigation strategy (Leung et al., 2002; Finnoff et al., 2007; Liebhold, 2012; Lodge et al., 2006), but we need to keep in mind that any prevention methods that involve raising the prices of imported goods imposes a cost on consumers as well as preventing IAS damages. The consumer is often left out of the prevention vs. eradication discussion altogether. This is partially due to imperfect international trade agreements not requiring consumer gains and losses to be considered. It is also a consequence of beneficiary industries being low in number and relatively well organised while consumers are large in number and largely disorganised.

Traditionally, both prevention and eradication have been viewed as the role of government regardless of the IAS involved. Where prevention is not achieved, most governments see their role as removing or eradicating introduced threats before the combined costs of impact and management grow to damage the economy. When eradication is not achieved and IAS becomes a permanent addition to a region after it has been introduced, responsibility usually shifts from government to the private sector. Only rarely would a government commit to paying the cost of continued management of established IAS.

This chapter demonstrates how economic analyses can play an important role in identifying the benefits and costs of alternative IAS risk-mitigation and outbreak response actions (i.e. prevention vs. eradication), and can therefore be used to distinguish between desirable and non-desirable policy options to improve resource allocation.

6.2 Cost–Benefit Analysis

In this section, we consider the pros and cons that must be weighed up by industry and government when deciding how best to respond to IAS risks, particularly when risk turns into reality following an incursion. No matter what form a response takes when an IAS is detected, the higher the net benefit it generates over its duration the more desirable it is for society.

Whenever a response involves the removal of an IAS from affected regions, it will generate a response benefit (RB). The benefits produced will depend on the extent and success of the removal process, which in turn will depend on the abundance and distribution of the IAS upon commencement of the response and what would have happened if no response had been mounted (i.e. the counterfactual). The investment of response effort (E) in a given response option will have associated with it a response cost (RC).

Response effort includes all labour and equipment purchased as part of a particular response option, and is subject to diminishing marginal benefits. That is, successive increases in response effort yield successively smaller benefits. If enough response effort is expended, the complete removal of an IAS can be achieved, at which point we might expect marginal response benefits to be at their lowest. However, this is not necessarily the case in every incursion response.

By calculating the difference between the present value of RB and RC for alternative response actions economists (and policymakers) can identify the most desirable option. This comparison between RB and RC is termed a cost–benefit analysis, or a benefit–cost analysis.

Cost–benefit analysis is the most widely used analytical decision support tool in biosecurity economics. It compares the expected present value of benefits generated by a particular policy to the expected present value of the costs of implementing that policy. The difference between these represents the likely net change in social welfare produced from the investment, or at least the component of social welfare determined by net financial gain.[6]

In its standard form, cost–benefit analysis indicates to policymakers which of a discrete set of policy alternatives is expected to produce the highest net benefit for society. Net benefits result from a surplus of benefits relative to costs, so by comparing them policymakers can establish the extent to which society is likely to be better or worse off under alternative biosecurity investments.

Assume we are informing a policymaker about a particular response or intervention characterised by the RB and RC curves shown in Figure 6.1. This diagram is static in the sense that it represents a single time period. Further assume that the intervention being considered involves the investment of response effort E_1 at a cost of RC_1. By removing a portion of the IAS population, benefits of RB_1 are generated from this response, which exceed RC_1. By calculating $RB_1 - RC_1$, we can indicate to the policymaker that this particular option would be expected to generate a net benefit of NB_1 for society.

If similar analyses are performed on different response options, a policymaker can be shown which option will generate the largest net benefit. Ideally, cost–benefit analyses will provide feedback into the actual design of policies themselves so that an option can be configured to deliver the maximum attainable net benefit. According to Figure 6.1, such a policy would involve an investment of effort corresponding to the point at which the marginal costs and marginal

---

[6]A second, less widely used tool is cost-effectiveness analysis in which only the costs of alternative means of achieving a defined objective are analysed, and generally the lowest cost policy alternative recommended. In this sub-section we will only concern ourselves with cost–benefit analysis. However, cost-effectiveness analysis can certainly be useful if there are difficulties observing response benefits.

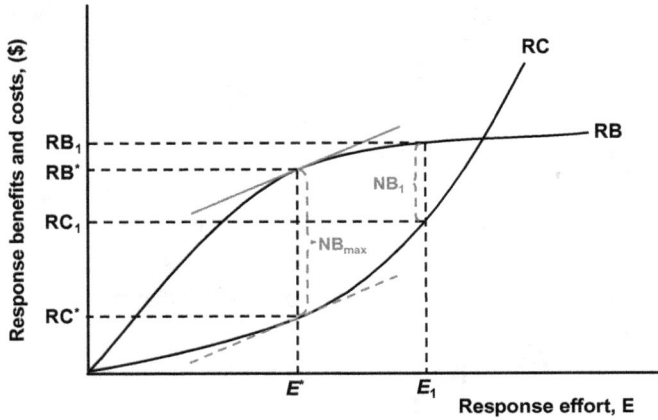

Figure 6.1. Cost–benefit analysis of an IAS response

benefits of response effort (i.e. the slopes of the RC and RB curves, respectively) are equal. Here, the cost of investing the last unit of response effort is exactly offset by the additional response benefit it generates. This occurs at E^*, which involves a response cost of RC^* and generates a benefit of RB^*. At this point the net benefits generated for society are NB_{max}.

In this particular example, the response involving a level of effort E_1 is inferior to one involving E^* since $NB_1 < NB_{max}$. Note however, unless a decision-support analysts is specifically asked to look at both response options, the policymakers they inform will have no way of knowing how an option like E_1 compares to E^* with respect to net response benefits. If they are informed of the option involving E_1, policymakers will learn that $RB_1 > RC_1$, and that resultant net benefits (NB_1) are positive. On the strength of such information, they might choose to invest in this option without ever knowing greater net benefits could have been created for society by investing a lower level of effort, E^*.

Whatever the option being considered, cost–benefit analysis is a very powerful way to show a policy's relative strengths to a policymaking audience. This is due to its ability to break down complex IAS impacts into one-dimensional financial indicators that are easily communicated. In fact, without cost–benefit analysis we would expect to encounter commensurability problems if we resorted to methods that express IAS impacts as lists of physical consequences. In the case of polyphagous (or multi-host) IAS, this could quite literally mean having to compare apples and oranges! Cost–benefit analysis avoids this problem by converting IAS impacts into easily-comparable monetary values.

There are no strict rules dictating how a cost–benefit analysis should be carried out, but we can identify a number of key steps in IAS-related analyses informing management decisions. They include the formation of:

1. Objectives — Determine the intended goal of an investment decision (e.g. to invest in the response strategy for an IAS that yields the highest net benefit to society);
2. Alternatives — Articulating a discrete set of investment options by which objectives might be achieved (e.g. prevention, eradication, slow-the-spread);
3. Benefits — Identify the benefits to be included in the assessment and the extent to which they are attributable to investment in each option (e.g. on-farm variable cost increments and total revenue losses);
4. Costs — Identify the costs involved in funding specific alternatives (e.g. capital and labour costs contributed by government and industry);
5. Spatial scale — Clearly identify the target area for the investment (e.g. a specific region or sub-region);
6. Timeframe — Establish the length of time over which the benefits and costs associated with each option are to be calculated (e.g. 10, 20 or 30 years);
7. Discount rate — Determine an appropriate rate of discounting through stakeholder consultation or recognised standard. The decisions should consider the extent to which social time preferences are relevant (i.e. dictating a lower discount rate, e.g. 2–3% per annum) or private time preferences (i.e. indicating a higher discount rate, e.g. 5–7% per annum);
8. Evaluation — Develop benefit and cost streams based on relevant data and information for each investment option. These should be

compared and reported using an agreed set of indicators (e.g. net present value, internal rate of return and/or benefit–cost ratio);

9. Decision — Ensure policymakers understand to the extent possible the information contained in indicators and choose an alternative. Where an alternative is not clear due to uncertainty in the evaluation phases of the analysis a sensitivity analysis can be performed to determine how different evaluation possibilities will affect the prioritisation of options. If a particular option is still preferred regardless of the evaluation values, the uncertainty is of little consequence to the policy decision. However, when the sensitivity reveals highly changeable preferences there may be a need for further research into the likely benefits and costs of different options.

Biosecurity decision support tends to involve *ex-post* (i.e. retrospective, or after the fact) cost–benefit analyses rather than *ex-ante* (i.e. before the fact) (Born *et al.*, 2005; Naylor, 2000; Cook *et al.*, 2010). Since the impacts of biosecurity decisions are often felt in time periods after an initial investment, *ex-ante* analyses can be important inputs to policy formulation processes. While *ex-post* studies assessing the effectiveness of biosecurity policies are highly challenging, they tend to be data rich when compared to *ex-ante* approaches. The latter requires researchers to predict how an IAS might be affected by increased (or decreased) biosecurity effort and how this will flow on to economic activities, the environment and society.

In the remainder of this chapter we will investigate how *ex-ante* cost–benefit analysis can be applied to IAS policy decisions related to the management of IAS risks using the model presented in Chapter 3. In particular, we will discuss how a decision-support analyst might carry out cost–benefit analyses of alternative activities that either prevent the arrival of exotic IAS, or eradicate them post-arrival.

6.3 Prevention

It is important to recognise that a 'zero risk' biosecurity policy is infeasible due to the sheer abundance and diversity of IAS, many of which are capable of reaching new regions by themselves without human assistance. Even a 'minimum risk' goal that accepts natural occurrences but not human-aided IAS arrivals is impractical as it would require bans on all passenger and freight movements into a region. While this would certainly limit damage from IAS, it would destroy the industries the policymakers were trying to protect as they would no longer be able to purchase necessary inputs to remain in production.

If we depart from zero and minimum risk to a 'managed risk' policy, we must accept the fact that IAS incursions will occur from time to time and that it is necessary to maintain a reactive capacity to respond to new species invasions as they happen. Any such response must aim to achieve the highest possible net benefit for society given the context in which it takes place, which means that economic analyses have an important role to play in managing both pre- and post-border IAS risks.

A pre-border prevention strategy reduces the probability of entry and establishment of an IAS. Preventative biosecurity policies for many IAS may entail inspection and interception at points of entry, such as docks and airports. Or they may involve pushing the reach of domestic biosecurity services well beyond the regional boundary being protected. In the case of the WA, this may involve the use of phytosanitary market entry requirements for imported IAS host products, quality assurance and produce certification schemes.

These measures will have the effect of reducing the probability of entry and establishment of the IAS. This in turn will shift the total cost curve for a given IAS downwards. The resultant response benefits created will persist over time, and can be calculated as the difference between the expected total costs with and without the investment in response effort. Where costs and benefits of prevention activities are considered over multiple time periods (i.e. unlike the static framework presented in Figure 6.1 which considers a single time period), RB and RC values predicted in future time periods must be discounted to their present values.

To illustrate a dynamic cost–benefit analysis for a prevention technology, let us use the example of parasitic mites threatening

European honey bees (*Apis mellifera*) in Australia. Honeybees themselves are not native to Australia but provide substantial (and free) pollination benefits for plant-based industries. The future of these pollination benefits is threatened by four mite species that are currently exotic to Australia: *Varroa destructor*, *V. jacobsoni*, *Tropilaelaps clareae* and *T. koenigerum*. Each species parasitize adult *A. mellifera* and their developing larvae, resulting in the death of wild honeybee colonies and potentially huge production losses for pollinator-dependent plant-based industries.

Using a similar model to the one described in Chapter 3, it is estimated the mites could cause $75 million per annum in damages through disruptions to wild pollination services to agriculture if they were introduced to Australia (Barry *et al.*, 2010). The mean total cost curve appears in Figure 6.2.

Now consider a hypothetical investment decision involving the potential enhancement of a sentinel hive network at ports and freight terminals that would increase the probability of detecting new mite-carrying honeybee incursions, thus lowering the probability of

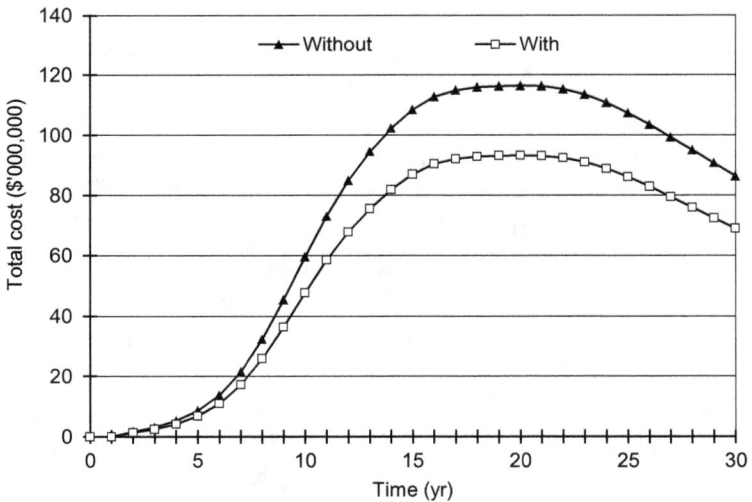

Figure 6.2. Preventing bee mite species from becoming established in Australia

each mite species becoming established. Assume that these measures could lower the probability of mite establishment such that the combined probability of entry and establishment (or the probability of arrival, z, from Chapter 3) falls by 10%.

Figure 6.2 shows the total cost curves associated with bee mite incursions over time with and without the proposed surveillance measures. The present value of total cost expected from bee mites without surveillance averages \$73.2 million per annum. This is an annualised representation of costs over the 30-year timeframe of the analysis (i.e. the total cost summed over the whole 30-year period is \$2.2 billion). With surveillance measures in place and a resultant lower probability of establishment, the total cost falls to \$65.9 million per annum (i.e. \$2.0 billion summed over 30 years). This implies that the proposed surveillance strategy will prevent approximately \$7.3 million plant industry costs per year from lost pollination services resulting from mite incursions (i.e. \$219.6 million over 30 years).

The predictions of total costs can be used in a speculative cost–benefit analysis that indicates to policymakers whether or not investing in an enhanced sentinel hive network is likely to produce a net benefit for society over time, and what the likely size of this net benefit (or cost)is. If the expected total cost of bee mites can be lowered to the extent indicated in Figure 6.2 with an investment of \$1.0 million per year, the ratio of benefits to costs is approximate 7.3:1.0. However, if a 10% reduction in the probability of entry and establishment is not achievable without the invest-ment of \$2.0 million per year, the ratio of benefits to costs falls to 3.7:1.0.

For our hypothetical reduction in the probability of establish-ment, an investment of \$7.2 million per year represents a 'break-even' point for policymakers where the ratio of benefits to costs would be 1:1. For investments over this value, the costs of the improved surveillance measures would outweigh the expected reduction in bee mite costs over time.

Box 14. What about Slowing-the-Spread?

Investments in technologies that could potentially slow-the-spread of an IAS post-arrival can be assessed in much the same way as preventative technologies. The key difference is that instead of lowering the probability of entry and establishment (z), slowing-the-spread influences the population/infection diffusion (D), intrinsic rate of affected area increase (r) and/or the intrinsic rate of satellite generation per unit area of infection/infestation (μ) parameters (i.e. see Table 3.1). However, the effect on the cost curve will essentially be the same as that depicted in Figure 6.1. Sticking with our Varroa example, it is predicted that inter-hive spread could be reduced by around 25% if research funds could be targeted towards the replication of hives with demonstrated grooming behaviours that remove the mites. This suggests a benefit in present value terms of approximately $6.25 per annum. Using a cost–benefit approach, the research would produce a net gain to society if it could achieve a 25% reduction in impact for less than $6.25 million per annum.

We can conclude that the role of cost–benefit analysis in pre-invasion risk management strategies is to weigh up the expected gains from utilising prevention technologies or tactics and the costs involved in developing and installing them. If these development and implementation costs outweigh the anticipated benefits (i.e. the area between the RC and RB curves in Figure 6.1), a cost–benefit cost analysis will reveal a benefit to cost ratio that is less than 1:1. On the other hand, if costs are less than expected benefits, then the prevention activity can be expected to (more than) pay for itself over time.

6.4 Eradication

6.4.1 *Static*

The complete removal of some IAS can be achieved with sufficient investment of biosecurity effort. When successful, the (re)attainment of IAS area freedom can produce non-localised benefits for host industries that extend to all growers, not just those that have had to cope with an IAS on their properties. If the eradication of an IAS means that export markets are restored after temporary bans, a sharp increase in RB is generated. This spike in benefits results from the sudden export revenue generated from the reclamation of area-freedom. Beyond the level of effort corresponding to eradication there are no additional benefits to be had by further investment in the response.

The total benefits from an IAS incursion response described above are depicted in Figure 6.3. Here, total benefits associated with a response option, RB, are shown as a function of response effort invested, E, where:

$$RB' > 0 \quad \text{and} \quad RB'' < 0. \tag{6.1}$$

Beginning at zero and moving left to right along the horizontal axis, the returns to response effort increase at a decreasing rate until the point E_{erad} is reached, corresponding to eradication. There are fewer and fewer affected areas as biosecurity effort increases, so it requires larger and larger effort to achieve a constant increase in RB. However, export market losses resulting from the initial detection of the IAS

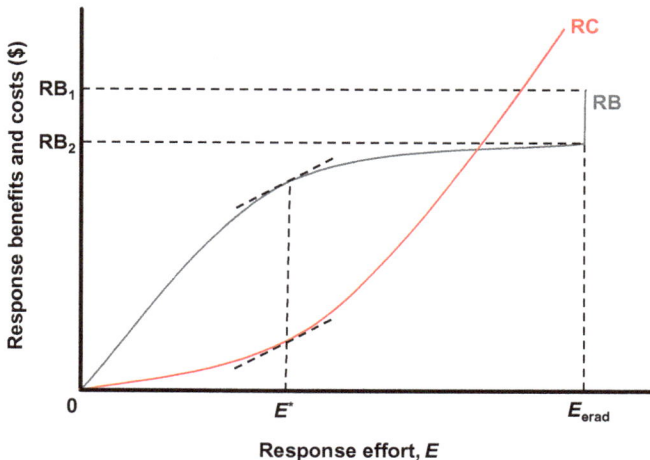

Figure 6.3. Total benefits and costs of an incursion response

remain until area freedom status is restored. Despite this requiring a high investment of biosecurity effort, E_{erad}, once reached the total response benefits suddenly increase to RB_1. This increase does not persist, and at levels of response effort beyond E_{erad} the returns will fall to zero. In contrast, when there is no export benefit to be gained from eradication, benefits are maximised at RB_2.

The extent to which $RB_1 > RB_2$ varies according to the IAS involved. Some IAS may cause no disruption to exports, some may cause limited export losses due to the need for costly pre-export requirements, and some could cause the complete prohibition of host products. In the case of plant IAS, export losses will usually be localised whereby movement of host material is prohibited within a specified distance of an IAS detection point, but is unaffected elsewhere. Exceptions certainly occur. As mentioned in Chapter 4, the market response to the wheat diseases like Ug99 could be severe (see also Thorne *et al.*, 2004; Brennan *et al.*, 1992; Stansbury and Pretorius, 2001).

In contrast to RB, RC is an increasing function of the level of biosecurity effort expended:

$$RC = c(E), \quad \text{where } c'(E) > 0 \quad \text{and} \quad c''(E) > 0. \qquad (6.2)$$

Beginning at zero and moving left to right along the horizontal axis, the costs of response effort increase at an increasing rate. As affected areas become fewer and fewer, more and more response effort and cost is required to achieve a given amount of response benefit.

Net benefits of a response option are maximised at the level of response effort that produces the greatest RB relative to RC. In the stylised example shown in Figure 6.1, this occurs at the response effort E^*, which is lower than the level of effort required to achieve eradication. This option involves 'slowing-the-spread' to an extent that would achieve some form of socio-environmental standard, rather than eradication *per se*. Hence, a response option involving E_{erad} does not represent a net-benefit-maximising solution.

The choice of whether to eradicate or control an outbreak will not always be as straightforward as in Figure 6.3. Consider an IAS with relatively large export market access implications, as depicted in Figure 6.4. Following an incursion of this type of IAS, as previously, policymakers need to decide whether to invest in a costly eradication response or a cheaper damage control (or slow-the-spread) option. According to the diagram, there would be nothing to choose between the two options if net benefits were the policymaker's only concern, as the net benefits associated with eradication (NB_1) are equal to those associated with slowing-the-spread (NB_2). If they also require the response option to be the most cost-effective (i.e. the lowest-cost), the 'slow-the-spread' option will be preferred and an investment of E^* will be made.

Box 15. Low-cost Eradication

On occasion, an IAS is detected early enough for the cost of eradication to be low. An example involves the invasive weed Mimosa (*Mimosa pigra*). This native of tropical America was introduced to Australia in the late 1800s as an ornamental plant; its appeal attributable to the curious property of the leaves folding when touched. Mimosa has formed dense thickets over more than 800 km^2 of tropical Australia (Paynter, 2005), but remained absent from WA until 2012 when approximately 100 ha of infested land was discovered in the tropical north of the State. The potential impact of the weed on cattle grazing enterprises was shown to be sufficiently large to warrant an eradication investment of up to $2.3 million. Given that eradication was deemed technically feasible for a cost of less than $10,000, the decision was made to eradicate the weed from WA.

6.4.2 *Dynamic*

If we consider Figures 6.3 and 6.4, it seems the only time E_{erad} would meet the criteria of a net-benefit-maximising response to an IAS would involve very low RC or a very high RB upon the restoration of export markets. However, by extending the time period over which an eradication response is evaluated, the decision of whether or not to invest in it becomes more complicated. It is no longer a matter of considering the benefits of removing a specific outbreak, but of

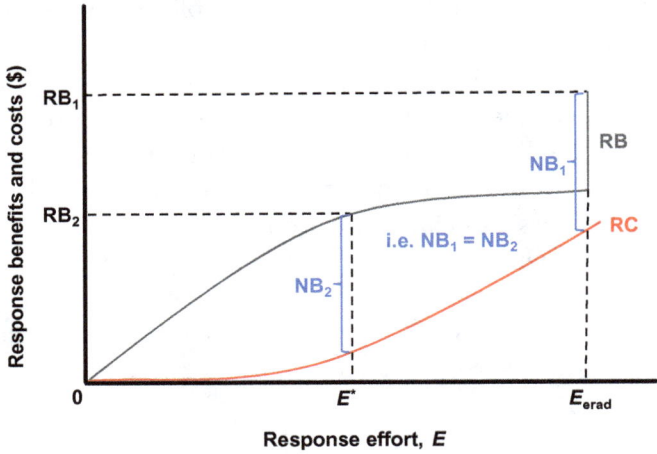

Figure 6.4.   Incursion response involving a large export component

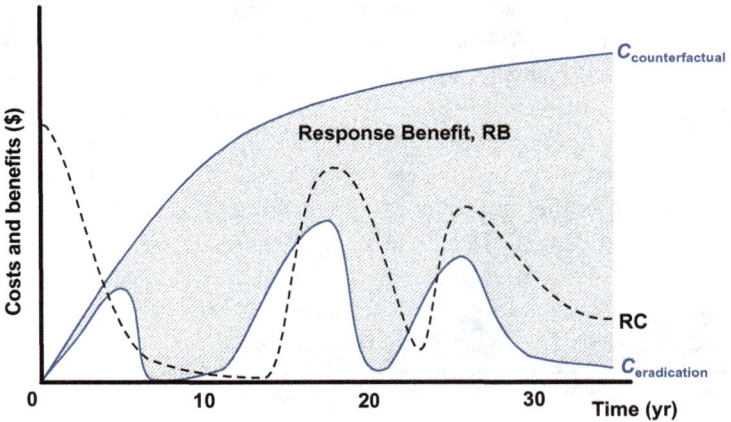

Figure 6.5.   *Ex-post* eradication cost-benefit benefits and costs over time

potential re-introduction events in future time periods given that the probability of arrival is greater than zero.

    Consider the situation depicted in Figure 6.5 in which we have plotted the costs associated with a hypothetical IAS over a 30-year period under two policy scenarios; one in which an eradication response is initiated upon detection in year zero (labelled $C_{eradication}$) and the other in which no response action is taken ($C_{counterfactual}$). The difference between them represents the RB.

Under the specific eradication scenario depicted, an initial knock-down is achieved between years 0 and 10 before the IAS population rises again, which could either be the result of another incursion event or a missed portion of the original incursion. This second event is suppressed but not completely removed between years 10 and 20, and again between years 20 and 30. The response costs involved are plotted by the broken line RC.

An *ex-post* cost–benefit analysis of this eradication attempt could be conducted to see if a net social gain was made, or if we might have been better off spending our scarce biosecurity resources elsewhere. It is too late to put these to other uses after the event, of course, but we could learn a valuable lesson for future incursion events. A retrospective cost–benefit analysis over the full 30-years eradication was attempted would subtract the present (or discounted) value of RC from the shaded region of Figure 6.5 (i.e. the vertical difference between the $C_{counterfactual}$ and $C_{eradication}$ lines in present value terms) to produce a net present value of the eradication policy. If this exceeds zero the cost–benefit analysis suggests eradication will result in a net positive social welfare result for society, and if not it suggests a net loss.

Without the benefit of hindsight, the initial policy decision of whether or not to attempt eradication requires a slightly different approach. If we are at time period zero and considering the future, an *ex-ante* cost–benefit analysis will help us to decide if eradication is worthwhile, but it will be highly uncertain. The $C_{eradication}$ line in Figure 6.5 is only one of an infinite number of possible future scenarios we need to consider. While we could conduct a cost–benefit analysis on this particular eradication scenario by defining the timeframe we are interested in, we have no way of knowing how plausible or how representative it is.

Indeed, if we could roll back time and create the exact same circumstances in which the initial incursion took place, there is no guarantee the $C_{eradication}$ would look the same as it does in the figure.

Ecologists tend to view environmental systems as characterised by complicated, chaotic interactions in which any outcome (stable or otherwise) is purely temporary (Farber, 2003). Natural variability is inevitable because of inherent biological stochasticity, so any observed IAS spread is only one replicate, or realisation, of a stochastic process (Melbourne and Hastings, 2009). Hence, our ability to forecast broad agro-environmental policy effects from IAS management effort is limited, at best.

Despite this being the case, policymakers can still be expected to ask for predictions to avoid nasty, politically-unpalatable surprises. Their success in dealing with agro-environmental change is partly determined by their capacity to anticipate consequences (Clark *et al.*, 2001). And, although we cannot hope to inform them with certainty with regard to IAS abundance and distribution over time, we can at least present realistic future scenarios using the stochastic model we developed in Chapters 2 and 3. Using the Monte Carlo method to simulate $C_{counterfactual}$ and $C_{eradication}$ many times, we can produce a probability distribution of each. This not only reveals mean, or expected values, but also variability around the mean.

For illustrative purposes, consider the mean $C_{counterfactual}$, $C_{eradication}$ and RC derived from multiple iterations of a model simulating the eradication policy described in Figure 6.5. These are shown as the 'expected' curves $E(C_{counterfactual})$, $E(C_{eradication})$ and $E(RC)$ in Figure 6.6, respectively.[7] We have also plotted the expected net benefit curve, $E(NB)$, which show the difference between costs and benefits through time. As the figure shows, response benefits are expected to be generated as soon as an eradication policy is put in place, but are initially too small to offset costs. Hence, expected net benefits are negative over the first decade or so, but steadily grow as the effects of the policy begin to take effect. As we near year 10 of the estimation period, the expected response benefits become large enough to offset the expected response costs, and thereafter expected net benefits are positive.

---

[7]Note that $E(RB) = E(C_{counterfactual}) - E(C_{eradication})$.

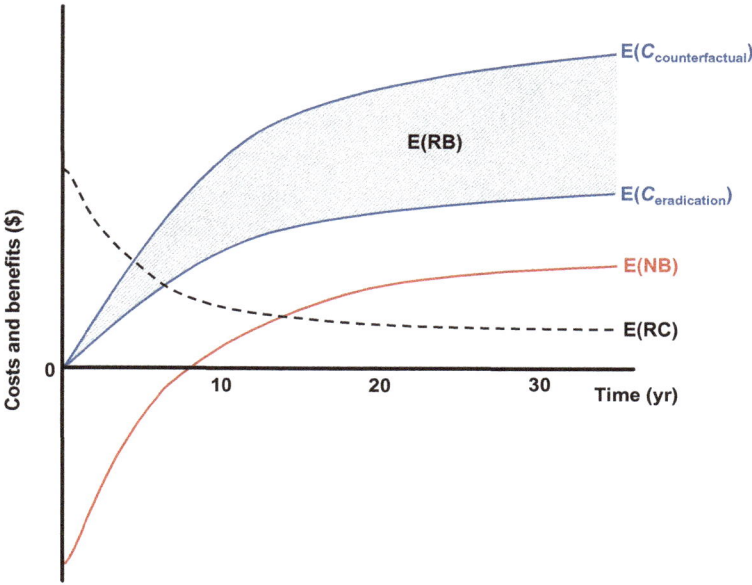

Figure 6.6. *Ex-ante* eradication cost-benefit benefits and costs over time

Assuming the timeframe over which we wish to examine the policy is 30 years, the net present value (NPV) of the policy (or the discounted value of the net benefit stream E(NB)) can be determined by summing the discounted expected net benefits in each year over 30-years. Effectively, the NPV provides policymakers with a single, comparable measure of the desirability of the policy:

$$\text{NPV} = \sum_{t=1}^{n} \frac{\text{E(NB)}_t}{(1+\alpha)^t}. \tag{6.3}$$

Here, α is the discount rate, n is the number of years we wish to estimate the NPV, and E(NB)_t is the expected net benefit resulting from the policy in time period t.

Consider an example of how this criterion can be used in conjunction with the stochastic model of Chapter 3 to decide between different IAS response policy options. Following the introduction of regulations restricting the use of organophosphate insecticides in WA to control Mediterranean fruit fly *Ceratitis capitata* [Wiedmemann], henceforth Medfly, industry and government began to consider an eradication policy. This pest has been established in WA for over 100 years. If relatively cheap and effective insecticides were no longer available, living with the pest would become more expensive for fruit growers as they would need to instead carry out intensive (and expensive) fly baiting and trapping regimes. A cost–benefit analysis was carried to determine if eradicating Medfly from the State would achieve a net benefit for the State (see Cook and Fraser, 2014).

Expected response costs (i.e. E(RC)) for an eradication program for Medfly were based on the use of the Sterile Insect Technique (SIT). Eradication was assumed to require a six-year program involving the release of approximately 100 million sterile male flies per week (Mumford *et al.*, 2001). The present value of costs this would involve was estimated to be \$94.2 million.

Expected response benefits (i.e. E(RB)) in this analysis were determined as the difference between costs accruing to fruit industries under an eradication scenario and a counterfactual scenario in which baiting and trapping was carried out by growers. When forming the eradication scenario, bear in mind that re-entry and establishment is likely to occur at some point or multiple points over the estimation period. Therefore, costs induced by Medfly under both scenarios calculated from multiple iterations of the model are positive, but as the timing of incursions across the temporal range simulated in the model is stochastic there is a large spread of possible spread scenarios.

These projections have been aggregated to produce Figure 6.7, which shows the expected costs to fruit growers if eradication (shaded bars) takes place and if a baiting and trapping (hollow bars) policy is adopted (i.e. $\mathrm{E}(C_{\mathrm{eradication}})$ and $\mathrm{E}(C_{\mathrm{counterfactual}})$, respectively) over a 20 years period.[8] All future values are discounted at 5% per annum

---

[8]The box whisker plot shown here is taken from Cook and Fraser (2014) and is the same as those used in the case studies of Chapter 4. They show the 25th, 50th and 75th percentile, and remaining values up to and including the 5th and 95th percentiles. The boundaries of the boxes indicate the 25th percentiles and the 75th percentiles, so the larger the box the more uncertain we are about our predictions. The horizontal lines inside the boxes represent the medians. Values lying between the 5th and 25th percentiles

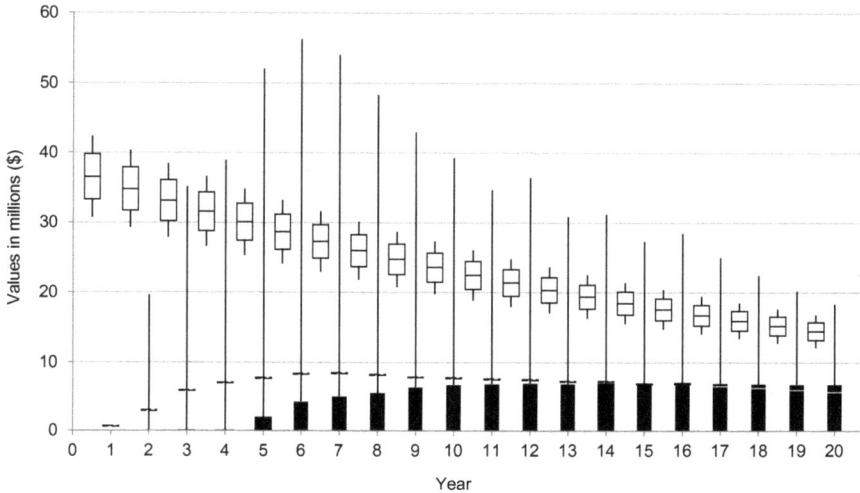

Figure 6.7. Predicted industry losses from Mediterranean fruit fly in WA over time

(Commonwealth of Australia, 2006). In the counterfactual scenario, Medfly remains present throughout the State and must be controlled to minimise fruit losses. The value of control costs is eroded through the process of discounting, so the impact falls in real terms over time. It is important to note that this is not due to a fall in pest prevalence over time.

If we summarise each cost flow as an average present value, a baiting and trapping approach is expected to cost the State approximately $528.0 million over 20 years, while eradication is expected to involve costs of around $132.0 million. Hence, the expected value of response benefits (i.e. E(RB)) were estimated to be approximately $396 million over 20 years. With expected response cost of $94.2 million over the same period, the NPV of adopting eradication over a baiting and trapping strategy was estimated to be $301.8 million over 20 years.

---

and between the 75th and 95th percentiles are shown by the lines extending from the top and bottom of each box.

6.5 Evaluating Prevention vs. Eradication Policy Options

Comparing the benefits and costs of a single policy option for a specific IAS is one thing, but what happens in cases where there are multiple policy options on the table? For instance, what if policymakers need to decide between investing scarce biosecurity funds on preventative technologies that lower the probability of an IAS incursion and eradication technologies that lower the cost of removing incursions when they occur? In such cases, decision-support analysts can use *ex-ante* cost–benefit analyses to compare the net benefits of each policy and help the decision maker to make a hard choice. In this section we will look specifically at prevention vs. eradication, but the general approach we use can just as easily be applied to slow-the-spread options.

To do this in general, we need to know the timeframe over which the resource allocation decision is to be made. Then, expressing each policy option in terms of the net benefits it is expected to produce over time allows a direct comparison between the alternative strategies. Moreover, this re-enforces the need for *ex-ante* analyses to estimate the RB and RC functions related to prevention and eradication for individual species. These analyses will also need to consider a variety of points along the cost-time curve (i.e. C, recalling Chapter 5) for each species to allow for time lags between arrival and detection.

In some cases, the comparison may be quite straightforward. If, for example, an IAS has proved extremely difficult to locate and remove after establishment elsewhere in the world, there may be no economic solution through eradication should it ever become established in a new region. That is, no matter what the investment in effort required to eradicate it, the costs to the region's economy will always outweigh the benefits of eradicating the IAS. Policymakers informed of this prior to an incursion would be inclined to invest more heavily in policies that prevent its entry into a region in the first place (e.g. pre-border phytosanitary measures), rather than policies that strengthen response capabilities (e.g. border protection, on-the-ground biosecurity personnel and equipment).

Other cases might involve more difficult policy comparisons. To illustrate, let us return to the example of bee mite IAS described in Section 6.3 (i.e. *V. destructor, V. jacobsoni, T. clareae* and *T. koenigerum*) and consider a situation in which a policymaker wants to know which policy option, prevention or eradication, is a better strategy in the long term to manage the risk associated with these species. We learned from Figure 6.2 that the present value of total costs to Australian plant industries from these mites changes considerably over a 30-year timeframe, but that on average the costs if we do nothing is expected to be around $73.2 million per year. This can be lowered by 10% if policymakers support a prevention policy involving sentinel hives. Assuming the program will cost $1.0 million per year in real (i.e. present value, or discounted) terms over 30 years and a discount rate of 5%, Figure 6.9 plots the flow of costs and benefits over time. Here response benefits are shown by the blue line RB, response costs by the broken line RC, and net benefits by the red line NB.

As in Figure 6.2, Figure 6.8 shows considerable variation in the flow of benefits from the prevention policy, making it somewhat difficult to summarise for a policymaker. Over the full 30-year timeframe shown in the diagram, the NPV is $156.5 million, implying an average net benefit of just over $5.2 million per year.

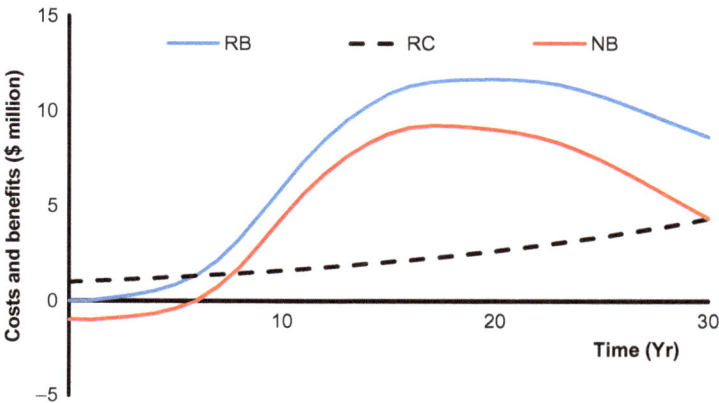

Figure 6.8. Hypothetical benefits and costs of a bee mite prevention policy

Suppose a policymaker wishes to compare this to an alternative policy of eradicating mites if and when they become established. Such a policy would be very risky in the case of bee mites; eradication might not be technically feasible if there is a sizeable lag between arrival and detection, and without a sentinel hive program they would be difficult to detect. However, in this hypothetical example we will use *V. destructor* eradication cost estimates from New Zealand of $70 million (Belton, 2000) as a proxy for eradication costs in Australia, as no equivalent eradication cost estimates for Australia have been put forward. Although more costly than the prevention policy, assume the eradication policy is expected to reduce the total cost of bee mites by 60%.

If a bee mite incursion takes place at an average frequency of one year in 5 over a 30-year period, Figure 6.9 provides indicative response benefit, response cost and net benefit lines (RB, RC and NB lines, respectively) we might expect from multiple iterations of a bioeconomic model. Note that eradication costs have been discounted over time, and so appear to decrease with successive time periods. Over the 30 years simulated by the model, the NPV of the eradication policy option is $134.4 million, implying an average net benefit of just under $4.5 million per year.

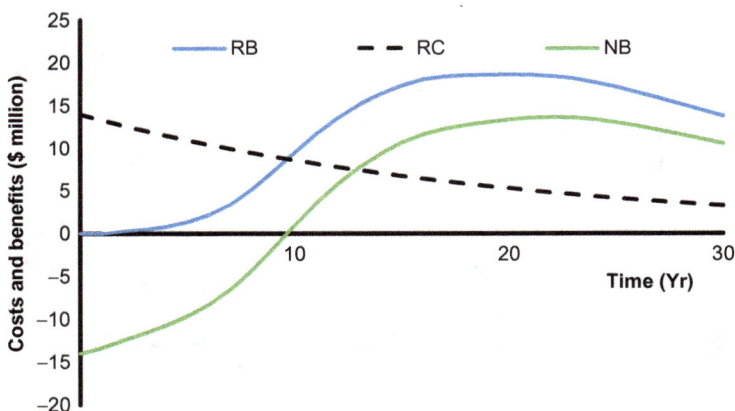

Figure 6.9. Hypothetical benefits and costs of a bee mite eradication policy

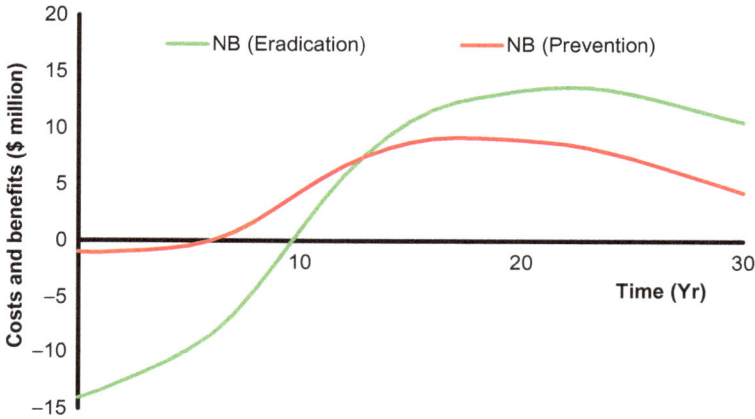

Figure 6.10. Hypothetical net benefits of bee mite eradication and prevention policies

If 30 years is indeed the timeframe the policymaker is interested in, the comparison between the two policies is relatively straightforward. The net benefit of a prevention policy is approximately $22.1 million more than what can be expected by an eradication policy. However, if we plot the net benefit lines of both policies together (Figure 6.10), we can see that the choice is more complicated. The net benefit of the eradication policy in the 30th year of the simulation is actually higher ($10.6 million) than that of the prevention policy ($4.3 million). Indeed, if we only consider the annual net benefits, eradication is the preferred option from year 14 of the simulation onwards. However, on average over the course of the full estimation period the policymaker can expect higher net benefits from investing in prevention.

In summary, this sort of economic decision rule, if used in a biosecurity resource allocation context, will enable a reasoned choice of how best to allocate a fixed budget between prevention and eradication/control policy options for the net benefit of society. However, specific cases will require specific analyses involving sensitivity testing.

In our example, the policymaker choosing between prevention and eradication for bee mite species would benefit from testing the sensitivity of the eradication alternative to changes in the probability of bee mites actually arriving and having to be eradicated over the 30 years of the assessment period. For instance, if an incursion event occurs one year in 10 (rather than one in five), response costs associated with the eradication option fall from an average of $7.4 million per year to $3.7 million. However, the likelihood of successfully eradicating them once they have been detected may be low, making the RB curve of Figure 6.9 a great deal more uncertain that it appears in the diagram.

Needless to say, the comparison between policies (be it prevention, eradication or a slow-the-spread option) needs to be made carefully. The bioeconomic modelling approach we have advocated in this book has the advantage of allowing broad comparisons between IAS management policies, but it is important that the uncertainties of both options are appreciated by decision-support analysts and, to the extent possible, communicated to policymakers.

6.6 Conclusion

In this chapter, we have seen how cost–benefit analysis can play an important role in policy decision support by summarising a wide range of complex information. This makes it an ideal tool to use when communicating with time-poor policymakers who simply lack the time to carry out the detailed research necessary to understand the nuances of each and every IAS policy decision. We have seen how cost–benefit analysis can be applied to response decisions to determine if prevention or eradication policies are preferable, and how dynamic *ex-ante* analyses are needed to help policymakers to think about future risks.

As powerful as cost–benefit analysis is, in the next chapter we will discuss problems that arise when we move away from taking account only of economic IAS impacts to also taking account of environmental and social impacts. Here, it can be much more difficult to elicit response benefits given that assets affected by IAS do not have market values that we can derive from statistical databases. Uncomfortable though this may be for decision-support analysts, IAS make no distinction between well-husbanded cultivated crops or rare and endangered environmental hosts. Their motivation is to survive by any means necessary, so our job as decision-support analysts requires a degree of plasticity in terms of the tools and techniques we use to guide biosecurity policies.

CHAPTER 7

DEALING WITH NON-MONETISED ENVIRONMENTAL AND SOCIAL IMPACTS

7.1 Introduction

As shown previously in this book, it is important that response efforts be subjected to economic evaluation to ensure resources are being put to net-benefit-maximising use. If not, situations may arise where great expense is incurred in managing an outbreak of an IAS of relatively minor significance whilst more serious outbreaks do not receive the attention they deserve. Cost–benefit analysis in its standard form indicates to policymakers which of a discrete set of biosecurity policy alternatives is expected to produce the highest net benefit for society.

When inadequate or insufficient information is available to complete a cost–benefit analysis, decision-support analysts can use other tools as adjuvants. Deliberative, or group-based multi-criteria evaluation techniques in particular acknowledge the severe information constraints often facing users of cost–benefit analysis, and rather than assume them away provide a feedback loop to policymakers and the community to openly explore solutions and compromises in the context of specific policy decisions. Although sometimes viewed as imperfect substitutes for conventional cost–benefit analyses (e.g. Dobes and Bennett, 2009), we take the view that deliberative multi-criteria methods are very much complementary to cost–benefit analyses.

In this chapter, we will highlight three information constraints that can stifle the effectiveness of cost–benefit analyses. These involve incomplete sets of benefits and costs being considered, difficulties in accurately valuing these benefits and costs, and when the uncertainty about these benefits and costs is not made explicit. We will then explore the use of deliberative multi-criteria decision analysis to bolster conventional cost–benefit analysis in the context of biosecurity policymaking. We will demonstrate the use of such techniques to prioritise IAS responses on the basis of impact where damages span economic (agricultural), environmental and social assets.

7.2 Problems Encountered When Using Traditional Cost–Benefit Analyses

7.2.1 *Coverage and consistency*

One of the principle strengths of cost–benefit analysis — the ability to condense complex problems into simple metrics that are easy to communicate to policymakers — can also lead to problems when applied to IAS issues, particularly when this involves broad spatial and temporal scales. When large, heterogeneous land areas are (potentially) affected by IAS incursions, a formidable information constraint confronts decision-support analysts trying to establish costs and benefits associated with impact-limitation or mitigation investment decision.

Direct market-based benefits and costs of the kind we have considered up to this point are in some ways the easiest to include in cost–benefit analyses, but accurate information is seldom available to form large-scale predictions an analyst can be sure of. It is perhaps unsurprising then that attempts to aggregate the economic costs of IAS across landscapes have tended to involve extrapolations from local to national scales, leading to widely varying cost estimates. For example, the costs of IAS to the United States economy from two studies, Office of Technology Assessment (1993) and Pimentel *et al.* (2005), differ by a factor of 2.

Cost–benefit analyses have often been used to ascertain the net gains or losses that could result from importing goods from international sources with known IAS, but they have not followed a consistent format. The way in which the economic implications of imports have been estimated has been assessed on a case-by-case basis, rather than using a standardised cost–benefit method. Case studies have used a variety of economic analyses, including those that simply assume an outbreak scenario only affecting producers, those that seek to put a probability on this occurrence, those considering both consumer and producer impacts, or combinations of these (Cook and Fraser, 2008).

Hinchy and Low (1990) addressed a New Zealand request made in 1989 to import apples into Australia, where the major disease transference concern was fire blight, a disease caused by the bacteria *Erwinia amylovora* that affects apples and pears. Australia's detailed response to this request included an economic component Hinchy and Low (1990) which took the form of a cost–benefit analysis comparing expected consumer and producer welfare changes resulting from relaxing quarantine laws protecting the apple industry. In 1995 New Zealand made another request to access the Australian apple market. This time the economic analysis, Bhati and Rees (1996), was quite different in approach. Expected consumer welfare change was omitted, and the analysis only considers possible producer surplus losses if a fire blight outbreak were to occur.

A market access application concerning salmon products from New Zealand, potentially forming a pathway for Whirling Disease of salmon, also prompted an analysis of economic consequences, McKelvie (1991), which uses a deterministic model. This analysis builds an entry scenario involving the introduction of whirling disease to three prominent Tasmanian fisheries and derives possible damage estimates. Neither the likelihood of disease arrival, the effect on domestic salmon consumers, nor the likelihood of scenario occurrence is discussed. Following a similar market access request from Canada a second economic analysis was prepared, McKelvie *et al.* (1994). This analysis dealt with two salmon diseases considered an importation risk. Again, the analysis comprises of a gross estimate of producer welfare loss in the event of a disease incursion, rather than a net welfare assessment (Cook and Fraser, 2008).

Applications by the United States, Denmark, Thailand and New Zealand to export chicken meat to Australia were the topic of another economic impact assessment. The potential economic implications of importing from these countries were examined in Hafi *et al.* (1994), which used one potentially imported disease (Newcastle disease) to illustrate the possible consequences of relaxing quarantine protocols. The method used in this analysis is similar to that of Hinchy and Low (1990) in that a critical probability of disease arrival is determined which brings the benefits and probable costs of trade into balance. Trade benefits were calculated as the change in consumer welfare resulting from lower domestic prices for chicken products, while the costs calculations were based on a severe Newcastle disease outbreak scenario causing a contraction in domestic supply of close to 20% (Cook and Fraser, 2008).

These assessments have been very useful as standalone decision support tools, but lack a common framework or accepted list of factors to include or exclude. The examples highlight an absence of consistency often encountered with cost–benefit analyses, which can make it difficult to compare different studies, or to draw meaningful conclusions from their comparison.

7.2.2 *Uncertainty*

Prioritising investment in IAS risk management activities is made very difficult by the uncertainty in arrival, spread and impact processes. It is important that wherever possible this uncertainty be taken into account by decision-support analysts and not assumed away through the use of deterministic models for the sake of brevity. This is easier said than done of course, particularly given time constraints placed on policy analyses, but if we look closely at uncertainty in biosecurity-related issues we can identify two general forms encountered when allocating our scarce biosecurity resources. The first, *epistemic uncertainty*, affects our ability to accurately measure important variables like the probability of arrival, intrinsic rate of affected area increase, population/infection diffusion coefficient, and so forth. The second perhaps less obvious (but no less pervasive) form of uncertainty is *linguistic uncertainty* arising from ambiguous use of terminology.

Simply put, epistemic uncertainty results from incomplete knowledge about the system under study (Merz and Thieken, 2005). There are many different sub-types, but these can be summarised as *natural uncertainty, pure uncertainty* and *ignorance*. Natural uncertainty results from underlying stochastic processes for which possible outcomes and their probabilities are known (e.g. throwing a dice or tossing a coin). Pure uncertainty describes instances where we only know the possible outcomes but not the probabilities of these outcomes (e.g. estimating wildlife reproductive rates where we cannot accurately predict the multitude of factors that affect them, but we do know the range over which reproduction is possible). Ignorance, or absolute uncertainty, occurs when we do not even know the range of possible outcomes (e.g. predicting ecosystem responses to climate change and human adaptability to resultant ecological change).

In the case of IAS, we often face a situation of ignorance in which we have great difficulty predicting whether any human actions will result in introduction, establishment and spread of an IAS, or whether a successful invader will have serious economic effects over time. For example, the red imported fire ant (*Solenopsis invicta*) is regarded as a serious threat to humans, agriculture and ecosystems across many regions of Australia following its accidental introduction in early-2001 (Scanlan and Vanderwoude, 2006); echoing concerns voiced after their introduction to the Southern United States in the 1930s (Vinson, 1997). Paradoxically, a decade on from fire ants invading the State of Texas they were regarded as a "benign presence" (Strayer *et al.*, 2006). Given such inconsistencies in anecdotal evidence, it can be difficult to characterise different policy options in cost–benefit analysis with any degree of confidence.

Linguistic uncertainty occurs as a result of difficulties in translating the causes and effects of IAS to those people or groups who are either impacted by or have the inclination and capacity to react to an IAS outbreak. Though often overlooked in cost–benefit analysis, linguistic uncertainty can be particularly pervasive in settings where expert opinion is sought using consistent terminology interpreted inconsistently by experts, resulting in misunderstanding and arbitrary disagreement (Carey and Burgman, 2008; Webb and Raffaelli, 2008). One familiar example is the potentially confusing set of terms developed around IAS, including non-indigenous, exotic, alien, pest and invasive species (Lodge *et al.*, 2006).

7.2.3 *Valuation*

The task of allocating resources to biosecurity is particularly complex in cases where prevention and response measures protect both market and non-market assets. Policy decisions determining who wins and who loses in these circumstances are multi-faceted, involving diverse groups of community and industry stakeholders with different priorities and objectives (Linkov *et al.*, 2004). In these cases, the economic analyses we have discussed so far using narrow market-based IAS cost calculations must be supplemented by other information; for they are only relevant in a policymaking context to the extent that market-based impacts are determinants of social welfare. As vague a concept as welfare is, we have a pretty good idea that changes in it are not solely determined by economic (agricultural) impacts. Decision support information must therefore consider a broader set of performance indicators when predicting the implications of policy.

Over the last quarter of a century, the valuation of environmental damage has received a great deal of attention from economists. The catalyst for much of this work came in 1989 when the oil tanker *Exxon Valdez* struck Bligh Reef in Prince William Sound, Alaska, spilling more than 11 million gallons of crude oil. The spill endangered millions of migratory shore birds and waterfowl, as well as many other species such as sea otters, porpoises, sea lions, and several whale species. In response to public outcry over an environmental catastrophe of such proportions the number of publications concerning environmental valuations jumped from less than 10 in 1990 to almost 100 by 1991, and by 2003 the number was over 470 (Adamowicz, 2004).

Box 16. Methods to Estimate Non-Use Values

There are numerous methods by which non-market goods can be valued, including:

Revealed preference approaches

— Travel cost: Valuations of site-based amenities are implied by the costs people incur to enjoy them (e.g. cleaner recreational lakes);
— Hedonic methods: The value of a service is implied by what people will be willing to pay for the service through purchases in related markets, such as housing markets (e.g. open-space amenities).

Stated preference approaches

— Choice experiments: People are asked to state their preferences for alternative policies, products or services described by a series of attributes, and responses used to test the statistical influenced of each attribute (e.g. choosing between remnant vegetation strategies with different levels of biological control effectiveness and cropping area);
— Contingent valuation: People are directly asked their willingness-to-pay or accept compensation for some change in supply of a non-market good (e.g. willingness-to-pay for preserved biodiversity);
— Conjoint analysis: People are asked to choose or rank different service scenarios or ecological conditions that differ in the mix of those conditions (e.g. choosing between wetlands scenarios with differing levels of flood protection and fishery yields).

Cost-based approaches

— Replacement cost: The loss of a natural system service is evaluated in terms of what it would cost to replace that service (e.g. using tertiary water treatment values as a proxy for the value of wetlands provide by reducing pollutants in streams and stormwater);
— Avoidance cost: A service is valued on the basis of costs avoided, or the extent to which it avoids costly averting behaviours, including mitigation (e.g. clean water reduces costly incidents of diarrhoea).

While much of this research has been effective in estimating *use* values for non-market goods derived from recreation and property value changes, it has been less successful in eliciting *non-use* benefits of environmental and social risk-mitigation (Adamowicz, 2004). These include existence, moral and bequest values contingent on the continued existence of the amenities at risk, which extend over generations in time (Mumford, 2001; Davidson, 2013).

These non-use values make valuation extremely difficult. The heterogeneity of preferences for non-use decision criteria — indeed the heterogeneity of specific non-market goods and their condition as we move from one resource allocation to another — present major obstacles for decision-support analysts. To date there is no practical means of eliciting non-use values for environmental goods cheaply, rapidly and accurately.

What we have instead are partial solutions; most notably *benefit transfer*. The benefit transfer technique adapts existing non-market value information or data to new policy contexts which have little or no data available from which to derive willingness-to-pay estimates. The transfer method involves obtaining a constant per unit estimate of the value of, for instance, an environmental or agricultural asset through a single study, or group of studies, that have been previously carried out to value similar assets in similar locations. The derived values from the original study site(s) are then transferred to another site of interest to policymakers. The per unit values can be adjusted to reflect local contexts using expert opinion (Costanza *et al.*, 2014), or a meta-analysis benefit function formed that statistically relates willingness-to-pay to characteristics of the ecosystems and communities involved.[9]

Proponents of benefit transfer seek to overcome the bottleneck of non-market valuation by making available databases of preference

---

[9] It is generally (but not universally) believed that unit value benefit transfer approaches are best suited to transfers between similar study sites and policy contexts, and that function transfers are more appropriate when the sites involved are dissimilar. However, there is no clear guide as to whether sites can be classified as similar or dissimilar, making it difficult to ascertain the relative performance of unit value and benefit function transfer approaches (Fetene *et al.*, 2014).

studies (Bergstrom and De Civita, 1999). Canada hosts one such international database that has been partly funded by the Australian government (see https://www.evri.ca/Global/Splash.aspx). Studies within these databases are either transferred directly to form a benefit function relating willingness-to-pay to characteristics of the ecosystem and the people whose values were elicited. When a benefit function is transferred, adjustments can be made for differences in these characteristics, thus allowing for more precision in transferring benefit estimates between contexts.

Extracting results from studies undertaken in one decision context and using it to imply willingness-to-pay in another is controversial (Morrison *et al.*, 2002). Problems arise because willingness-to-pay to protect specific environmental or social assets from a specific IAS at a specific point in time depends on an array of sociological factors. Many factors contribute to individual willingness-to-pay in different ways. While income exerts a positive influence (Whitby, 2000), for instance, other factors such as age have a negative effect (Saayman *et al.*, 2016). The willingness of populations to pay to prevent IAS biodiversity or ecosystem service impacts is therefore ambiguous, and it may take a proportionately greater amount of damage before a negative effect is perceived when compared to an IAS affecting agriculture.

This phenomenon has been seen in research attempting to extract willingness-to-pay values contingent on the state of environmental deterioration. For example, Blamey *et al.* (2000) asked survey respondents to distinguish between "non-threatened" and "endangered" species in stating valuations, while Hanley *et al.* (2003) evaluated respondents' views on protecting "all goose species" compared with "endangered goose species". In both cases the findings support an exponential dependence of social valuations of environmental goods on the extent of damage to those goods. It follows that the social valuation of an environmental good is likely to be not just positively but also exponentially related to the time path of its deterioration, suggesting a more inclusive set of investment criteria are needed to guide IAS investment decisions (Cook *et al.*, 2011b).

7.3 Facilitating IAS Policy Decisions Involving Market and Non-Market Effects

7.3.1 *Deliberative multi-criteria decision analysis (MCDA)*

In the remainder of this chapter we will discuss multi-criteria decision analysis (MCDA) and its role as a supplement to traditional cost–benefit analysis in IAS policy. MCDA is a decision-facilitation method, rather than a valuation tool. It has been developed as a means of simplifying complex policy problems involving multiple stakeholders, uncertain outcomes and a range of incompatible criteria by which to assess success or failure (Proctor and Drechsler, 2006). MCDA is particularly useful in the context of IAS policymaking as it is a relatively simple process by which we can establish public beneficiaries as well as private with a degree of expediency.

MCDA works by ranking different policy alternatives and identifying the option that has the greatest potential benefit for society. The rank order of choices are determined according to their performance against criteria that are weighted by stakeholders (Munda *et al.*, 1994; Rauschmayer and Wittmer, 2006). The weights applied to criteria reflect their relative importance in the eyes of stakeholders, but they are only useful when the options being considered score differently against them. If all options score the same against a criterion it can be discarded as it does not help us choose between them. Using remaining criteria in which scores are different, the scores for each option are aggregated into a single rank order.

In the deliberative form of MCDA we discuss, the rank order is presented back to stakeholders and a dialogue established to reveal the reasons behind the resultant hierarchy of policy alternatives. Re-weighting and sensitivity analysis are used to test the resilience of the rank order before a final decision on the best course of policy action is taken. The eventual policy choice represents the option that the stakeholders prefer given the context and time.

A wide variety of MCDA methods have been put forward for many different complex environmental and social decisions. They have been used to rank options, identify single alternatives, sort alternatives into groups, provide an incomplete ranking, or differentiate between acceptable and unacceptable policy options (Roy, 1985; Linkov *et al.*, 2004). Of the many MCDA variants, there is a growing trend toward the use of participatory methods, particularly within government, to create a more democratic and open process of funding allocation.

Participatory MCDA methods combine the facilitation, interaction, and consensus-building features of a jury process (Dienel and Renn, 1995; Crosby, 1999), with the structuring of conventional MCDA (Munda *et al.*, 1994; Massam, 1988; Proctor, 2005). The jury involves a small number of participants (i.e. between 10 and 20) assuming responsibility of constituent representation and decision making. The group is presided over by an independent facilitator who ensures that participants have equal input into the decision. Jury members can call on expert witnesses, technical analyses and anecdotal information to help form their individual opinions, and then time is devoted to group discussion, information clarification, and debate in which group opinions are revealed.

7.3.2 *Key steps in deliberative MCDA*

i) Choosing the IAS and the objectives of the analysis

The choice of options and objectives are closely related steps in any decision-making process. When prioritising IAS, individual IAS form the options, but we can also prioritise management strategies for specific IAS by using management alternatives as the options we want to choose between. The objectives and options can be chosen by the policymaker according to their needs and expectations, with potential input from experts, computer simulation models or political and/or legislative prerequisites.

ii) Jury selection

MCDA jury composition should be determined by an objective analytical process such as stakeholder mapping. This involves gauging a prospective juror's ability to influence the decision by providing or withholding resources, as well as the importance of the decision to them in relation to political, financial, social, environmental or technical interests. Classifying prospective jury members in this way provides an insight into potential coalition-building to increase influence. For example, members from different grower groups could combine to exert stronger influence in an MCDA than they could individually to emphasis a preferred IAS strategy.

iii) Criteria selection

The criteria are designed to compare and assess each of the options and therefore must relate to the overall objective of the policymaking task. Initially, criteria can be broadly defined and then broken down into components or sub-criteria and even lower level sub-criteria. Ideally, the lowest level of the criteria structure are those which are measurable (quantitatively or qualitatively) and are commonly referred to as indicators.

In general, ideal MCDA criteria will exhibit the following properties:

1. Clearly defined — Criteria should be free of linguistic uncertainty, at least from the jury's perspective;
2. Complete — Criteria should include all aspects of the policy decision on the economy, environment and society;
3. Mutually exclusive — Option scores relating to specific effects can be assigned independently from one criterion to the next without double-counting;
4. Divisible — Criteria can be divided into smaller units that are measurable;
5. Manageable — Criteria are not so numerous as to make the group deliberation component of the MCDA untenable. To allow adequate understanding of the decision problem and to assist in achieving a smoothly run process, a set of seven to 12 criteria is most commonly used (Hajkowicz and Collins, 2007).

iv) Criteria weighting

Juror preferences for desirable option characteristics are accounted for by weights placed on each criterion. The process of assigning weights can be divided into direct and indirect methods. *Direct* weighting methods estimate the exact worth of one criterion compared to another, whereas *indirect estimation* techniques use information on weights used in past studies. The particular weighting method chosen should be appropriate for the particular decision makers involved. For instance, some juries might be more comfortable using a ranking method rather than assigning weights directly on each of the criterion.

v) Evaluation of the options

Options are assessed by means of an *impact matrix* where the impact of each is scored according to criteria. In general, if one option performs better than another for all of the criteria then that option will be ranked highest. If the performance varies for different criteria (i.e. one option performs better for some but not all criteria in comparison to another option), then its rank will depend on how highly the superior performing criteria are weighted by the jury. The final ranking of each of the options is calculated by a mathematical operation. Growth in the field of MCDA has given rise to many different aggregation methods, ranging from simple weighted averages to sophisticated pairwise comparisons (Mendoza and Martins, 2006).

vi) Sensitivity analysis

Although not always undertaken in MCDAs that involve choosing a single optimal option, sensitivity analysis of the results can be a useful part of the process. It can be conducted in order to take into account uncertainty in estimation of values or weightings and may provide a range that can be statistically analysed. It can also consider the effects of different techniques used in the weighting procedure, or for different values of the most crucial and contentious criteria, weightings and impacts. For example, in a group decision-making situation, if it were found that there was a great disparity in preferences for a certain criterion that reflects a crucial trade-off, then it may be enlightening to find out how the overall results change when the weight applied to this criterion changes. If the results are not greatly affected, then the criterion can take less importance in the overall process and the jury can concentrate on other criteria. If the results are sensitive to this criterion, then closer scrutiny should be given to it by confirming values and measurements.

vii) Iterating and fine-tuning

The decision-support analyst can achieve greater understanding of the decision-making problem by interacting with policymakers to allow further iterations in the analysis if necessary, to identify where trade-offs can be made to concentrate on the important issues in the process and finally to re-define criteria and options to take into account what has been learnt from the process. This step can be important if the ultimate aim of the MCDA is to reach a compromise or agreement on the outcome.

While decision-support analysts may be familiar with uncertainty, policymakers and the public often accept scientific projections as certain. A management decision that assumes risk assessment results are certain, when in fact they are not, can result in unexpected or undesirable outcomes (Peterson *et al.*, 2003).[10] We contend that, like environmental policy support, IAS policy support is most effective when uncertainty is incorporated as information for hypothesis-building, experimentation, and decision making (Bradshaw and Borchers, 2000). In this

---

[10]In fact, the consideration of uncertainty may lead to a different decision in managing IAS risks (Burgman *et al.*, 1999; Regan *et al.*, 2005) Horan *et al.* (2002), for instance, argue that decision models based on standard economic theory have limited value when neither the range of potential impacts nor the possibility of these impacts is known for IAS management. They develop a model where policymakers cease maximising their utility and became uncertainty-averse instead. As a result, it becomes optimal to devote more resources to confronting high-impact events even if the probability is considered low.

sense, deliberative MCDA can be used to reduce policymakers' and stakeholders' discomfort about uncertain IAS management decisions.

7.3.3 *Case study — invertebrate response policy prioritisation*

In this section we present a case study applying a deliberative form of MCDA to facilitate prioritisation of policy choice related to an IAS as a demonstration case for deliberative MCDA methods. In addition, we explicitly incorporate uncertainty using a fuzzy set approach (Zadeh, 1965).[11] By integrating quantification and communication of uncertainties in this way, we can form a methodology that transparently incorporates epistemic and linguistic uncertainty rather than assuming them away.

Our case study involves a deliberative MCDA exercise conducted to determine an appropriate regulatory response from the building industry in WA to the threat posed by the IAS European House Borer (EHB); a timber pest that can cause severe structural damage to wood-framed buildings. Although this case study involved a real IAS, a real set of alternative regulatory options and real community stakeholders it was not directly tied to a real response. The exercise undertaken was designed to trial the effectiveness of the deliberative MCDA technique as a means of resolving IAS management issues, and as a means of showcasing the methodology to a group of interested parties.

For those readers more interested in the MCDA approach as means of incorporating and communicating uncertainties in policy decisions, we suggest skipping the following section and re-joining in Section 7.3.3.2. For those interested in the use of fuzzy sets to embed uncertainty in the decision-making process, Section 7.3.3.1 provides

---

[11]Fuzzy set theory was initially combined with MCDA in (Bellman and Zadeh, 1970). Due to its intuitive and computational ease of analysis, the fuzzy set approach has become one of the most common methods for dealing with uncertainty in MCDA (Kangas and Kangas, 2004; Liu *et al.*, 2010).

a brief overview of a suggested approach based on Liu *et al.* (2010), and earlier work by Yeh *et al.* (2000).

7.3.3.1 *Fuzzy logic*

The IAS management strategy performance evaluation problem involves a number (n) of discrete management options M_i $(i = 1, 2, \ldots, n)$. These options are to be scored according to a set of m criteria C_j $(j = 1, 2, \ldots, m)$, each of which is separated into p_j sub-criteria $C_{ik}(k = 1, 2, \ldots, p_j)$ (Yeh *et al.*, 2000). Since some criteria dealing with non-market IAS impacts cannot be scored with the aid of quantitative information, qualitative assessments can be provided according to a set of linguistic terms with corresponding fuzzy membership functions. Separate linguistic terms are used to determine: (a) appropriate scores for each management option against each sub-criterion, and; (b) the relative importance of each sub-criterion in choosing the most appropriate management strategy (Yeh *et al.*, 2000).

The linguistic terms defined in Tables 7.1 and 7.2 are used, each of which corresponds to a triangular fuzzy number representing their approximate value range between 1 and 9 (Juang and Lee, 1991).

Table 7.1. Linguistic terms used to seed the impact matrix

Qualitative term	Very poor (VP)	Poor (P)	Fair (F)	Good (G)	Very good (VG)
Membership function (a_1, a_2, a_3)	$(1, 1, 3)$	$(1, 3, 5)$	$(3, 5, 7)$	$(5, 7, 9)$	$(7, 9, 9)$

(Yeh *et al.*, 2000)

Table 7.2. Linguistic terms used to weight sub-criteria

Qualitative term	Least	Less	Fair	More	Most
Membership function (a_1, a_2, a_3)	$(1, 1, 3)$	$(1, 3, 5)$	$(3, 5, 7)$	$(5, 7, 9)$	$(7, 9, 9)$

(Yeh *et al.*, 2000)

The range is defined as (a_1, a_2, a_3), where $1 \leq a_1, 1 \leq a_2 \leq a_3 \leq 9$. The values of a_1 and a_3 represent the lower and upper bounds of the fuzzy number, respectively, while a_2 is the most likely value of a linguistic term (Yeh *et al.*, 2000).

The impact matrix (i.e. recalling step (v), above) can be expressed as:

$$X = \begin{bmatrix} x_{11} & \cdots & x_{1m} \\ \vdots & \ddots & \vdots \\ x_{n1} & \cdots & x_{nm} \end{bmatrix}. \tag{7.1}$$

Here, x_{ij} indicates the linguistic scores for IAS management option $M_i(i = 1, 2\ldots, n)$ with respect to criterion C_j $(j = 1, 2, \ldots, m)$ (Chang and Yeh, 2002; Yeh *et al.*, 2000). Since sub-criteria $C_{ik}(k = 1, 2, \ldots, p_j)$ are used for each criterion a more detailed impact matrix can be expressed as:

$$Y_{C_j} = \begin{bmatrix} y_{11} & \cdots & y_{n1} \\ \vdots & \ddots & \vdots \\ y_{1p_j} & \cdots & y_{np_j} \end{bmatrix}. \tag{7.2}$$

Here, y_{ik} are the jury's linguistic scores for the performance of management option M_i $(i = 1, 2, \ldots, n)$ with respect to sub-criterion $C_{ik}(k = 1, 2, \ldots, p_j)$(Yeh *et al.*, 2000). To truncate the prioritisation process a single consensus impact matrix was used in this study.

A weighting vector W_j $(j = 1, 2, \ldots, m)$ for the perceived importance of sub-criteria to a decision maker in making an IAS management decision is revealed using the linguistic terms in Table 7.3, and is expressed:

$$W_j = \left(w_{j1}, w_{j2}, \ldots, w_{jk}, \ldots, w_{jp_j} \right). \tag{7.3}$$

Here, w_{jk} are the fuzzy weights for sub-criteria $C_{ik}(k = 1, 2, \ldots, p_j)$. Sub-criteria weighting vectors can be elicited individually for each jury member. By combining the scored for each management alternative against each sub-criterion with the sub-criteria weights from one or more rounds of weighting by a jury, each alternative can be ranked in order of preference. We employ the widely-used concept of the *degree of optimality* to establish clear and defined preferences. The optimal management alternative is the one that is both closest to the ideal solution and farthest from the negative ideal solution (Zeleny, 1982; Benitez *et al.*, 2007).

The first step in the ranking procedure involves the formation of a weighted fuzzy impact matrix through the multiplication of the criteria impact matrix X (i.e. Equation (7.1)) with a consensus weighting vector W_j (i.e. Equation (7.3)) (Yeh *et al.*, 2000). A normalised preference function $(x_{1j}, x_{2j}, \ldots, x_{nj})$ for criterion C_j with sub-criteria C_{jk} is given by

$$(x_{1j}, x_{2j}, \ldots, x_{nj}) = \frac{W_j, Y_{Cj}}{\sum_{k=1}^{p_j} w_{j,k}}. \tag{7.4}$$

From the scores and sub-criteria weights expressed by a stakeholder jury, we can identify the IAS management alternatives with the maximum fuzzy preference value (M_{max}^k) and the minimum fuzzy preference value (M_{min}^k) with respect to each sub-criterion. The degree of preference for a single alternative with respect to each sub-criterion can then be calculated by comparing its weighted fuzzy performance with both M_{max}^k and M_{min}^k, i.e.:

$$\mu_{M_{max}^k}(x) = \begin{cases} \dfrac{x - x_{min}^j}{x_{max}^j - x_{min}^j}, & x_{min}^j \leq x \leq x_{max}^j. \\ 0, & \text{otherwise} \end{cases} \tag{7.5}$$

$$\mu_{M_{min}^k}(x) = \begin{cases} \dfrac{x_{max}^j - x}{x_{max}^j - x_{min}^j}, & x_{min}^j \leq x \leq x_{max}^j. \\ 0, & \text{otherwise} \end{cases} \tag{7.6}$$

Here, $j = 1, 2, \ldots, m$ and $k = 1, 2, \ldots, p_j$. Expression (7.5) gives a performance rating for each management option for each criterion based on how far it is from the criterion-specific most preferable option. This reflects the status of any one management option relative to the management option with the highest combined weight and score. This measure of preference is appropriate where the attitudes of policymakers to risk and uncertainty are "optimistic" (Yeh and Chang, 2009). In other words, the jury prefers the management option that most consistently exhibits a combined weight and score near the highest total for each criterion.

Conversely, Expression (7.6) gives a pessimistic preference weighting in which the jury seeks the option furthest from the worst option (Chang and Yeh, 2002). It is appropriate where the stakeholder jury seeks the management option closest to the best according to each of the sub-criteria. This is the concept we will use in the case study that follows, but we note it is also possible to form a pessimistic preference rating in which a jury seeks the option furthest from the worst option.

Equations (7.5) and (7.6) represent the preference limits for a jury. We represent this range of possible risk attitudes using an index λ with a value 0 or 1. $\lambda = 0$ and $\lambda = 1$ represent the optimistic or pessimistic view of jury members, respectively (Yeh *et al.*, 2000). An optimistic jury member prefers higher values of the fuzzy sets, while a pessimistic member is more conscious of low values. So, in the context of the case study presented in this chapter, an optimistic jury member pays more attention to management alternatives with higher scores, while a pessimistic jury member is more concerned about lower scores (Chang and Yeh, 2002).

By summing performance ratings $\mu_{M_{\max}^k}(x)$ and $\mu_{M_{\min}^k}(x)$ across sub-criteria for each IAS management option we can form an index of the degree of similarity S_j between each option M_i and the fuzzy maximum and minimum, $M_{\max}^k$ and $M_{\min}^k$, for each criterion C_j.

$$S_j = \lambda \sum_{k=1}^{p_j} \mu_{M_{\max}^j}(x) + (1 - \lambda) \sum_{k=1}^{p_j} \mu_{M_{\min}^j}(x). \qquad (7.7)$$

Here, $j = 1, 2, \ldots, m$. The higher the value of S_j the more preferable the management alternative is to the jury with a mixture of optimistic and pessimistic attitudes with respect to the IAS. The positive and negative ideal management strategy for the jury (i.e. the positive and negative extremes of risk preference, r^+ and r^-, respectively) can be stated as:

$$r^+ = \max \sum_{k=1}^{p_j} \mu_{M_{\max}^j}(x). \qquad (7.8)$$

$$r^- = \max \sum_{k=1}^{p_j} \mu_{M_{\min}^j}(x). \tag{7.9}$$

In order to determine the overall performance of each management option the index S_j (i.e. from Equation (7.7)) must be compared to both the ideal minimum and maximum. The difference between S_j, r^+ and r^- is referred to as the Hamming distance (Chang and Yeh, 2002), and can be calculated as:

$$h^+ = \sum_{j=1}^{m} \left(r^+ - S_{M_j} \right). \tag{7.10}$$

$$h^- = \sum_{j=1}^{m} \left(S_{M_j} - r^- \right). \tag{7.11}$$

A crisp overall index of the relative preference for any management strategy M_i is then calculated as:

$$P_i = \frac{h_i^-}{h_i^+ + h_i^-}. \tag{7.12}$$

In the case study which follows P_i is expressed in percentage form (see Table 7.7).

7.3.3.2 *European house borer in Western Australia*

EHB was discovered on the outskirts of the city of Perth in January 2004, but may have been present for 10 or more years prior to detection. This destructive IAS of seasoned coniferous timber affects pine, fir and spruce, and is therefore capable of causing structural damage to buildings. Residences with untreated radiata pine, southern pine, Douglas fir, hoop pine or bunya pine frames are particularly at risk.

Despite extensive surveying, at the time of the study EHB had only been found in several outer suburbs of Perth. A national cost-shared control program formally commenced in January 2007 under the Primary Industries Standing Committee[12] arrangement in which the Commonwealth government paid 50% of the costs of eradication while the remainder was funded by the WA State government.

To supplement this eradication campaign the WA Department of Housing and Works (DHW) commissioned a report investigating possible regulatory actions that could be taken to mitigate the impacts of the insect on the housing industry (The Allen Consulting Group, 2006). This report was intended to stimulate public discussions regarding possible regulatory options, but by November 2008 no regulatory actions had been decided upon. The three management alternatives put forward for consideration in the report were as follows:

(a) Do nothing

Under this scenario there are no additional building regulations put in place to guard against possible EHB damage. Hence, private home and business owners deal with effects of the insect in their own way;

(b) Statewide Building Restrictions

All use of untreated softwood building materials in new homes and businesses will be restricted. Regulations would be put in place banning the use of untreated softwoods for structural purposes ensuring that houses are structurally protected from EHB infestation. The proposed regulations would be monitored and enforced by local government authorities as part of the process which they currently undertake in granting approval to building applications;

(c) Delimited Building Restrictions

The use of untreated softwood building materials in new homes and businesses will be restricted in areas where the Borer has been detected. Hence, the structural quality of new homes in affected areas will remain at a safe level while limiting compliance costs faced by the community relative to a statewide approach.

---

[12]The Primary Industry Standing Committee arrangement pre-dated the Government and Plant Industry Cost Sharing Deed in Respect of Emergency Plant Pest Responses (Plant Health Australia, 2005) described in Box 9.

This alternative relies heavily on the assumption that EHB will not spread significantly beyond its current distribution.

The objective of the deliberative MCDA study was to evaluate these three regulatory options using a list of criteria developed in consultation with a stakeholder jury, and to decide on the most desirable option according to these criteria. The economic criteria are informed by The Allen Consulting Group (2006), which takes the form of a conventional cost–benefit analysis.

The jury was made up representatives from various community and business groups potentially affected by EHB. Approaches were made to local shire councils who would be in a position to enforce any building regulations agreed to; the State department of Housing and Works who commissioned The Allen Consulting Group (2006); the Department of Agriculture and Food, Western Australia; building industry associations; State and local environmental conservation groups concerned with insecticide usage. The jury on the day of the workshop totalled 10 individuals. In the results to follow the identity of jury members and their corresponding preferences is kept confidential.

The deliberative MCDA exercise was presided over by a professional facilitator. Through investigations and consultation with jury members, three criteria (*economic, social* and *environmental*) were identified as being vital to a prioritisation process incorporating a range of sub-criteria. These included three economic, four social and one environmental sub-criterion. The eight sub-criteria were defined by the jury as follows.

Economic criteria

(1) *Compliance costs*

The costs of adhering to the proposed regulations. These costs will ultimately be passed on to new home buyers through increased prices for susceptible building materials (The Allen Consulting Group, 2006).

(2) *Expected damage costs*

The damage costs of infested houses given the continuation of current EHB Program monitoring and containment activities.

Table 7.3. Quantitative assessment data and corresponding qualitative assessment results for the *Economic* Criterion

Sub-criterion	Management alternative		
	Do nothing	Statewide building restrictions	Delimited building restrictions
Compliance costs	$ -	$ 697,000,000	$ 37,000,000
	VG	VP	F
Expected damage costs	$ 120,000,000	$ -	$ 1,000,000
	VP	VG	F
Administration cost	$ -	$ 52,000	$ 52,000
	VG	G	G

(3) *Administrative costs*

Administrative costs include items associated with the design and implementation of regulations. In the case of the proposed regulations, administrative costs will accrue to the State government in terms of the initial implementation of the regulations (i.e. amendments to existing legislation) and local government in terms of monitoring, enforcement and the assurance of compliance.

Table 7.3 shows quantitative data from The Allen Consulting Group (2006) relevant to each of the economic criteria, and the jury's corresponding score for each management alternative using the scale defined in Table 7.1.

Social criteria

(4) *Reduction in infested houses*

This sub-criterion viewed in a social context takes into consideration the health and safety aspects associated with EHB infestations. Banning the use of untreated softwoods that are used for structural purposes will ensure that houses built after the regulations are implemented are structurally protected from Borer infestation. The regulations would protect the health and safety of: (i) occupiers of new houses (i.e. the regulations will protect against structural roof collapse due to infestation and also ensure that new houses maintain structurally strong enough to withstand extreme climatic conditions such as wind gusts); and (ii) those working on new houses (i.e. workers such as electricians or roof tilers that are commonly required to work within roof spaces or on top or roofs will be protected from suffering injuries due to structural collapse caused by infestation when working on homes built with treated structural timbers).

It is difficult to accurately quantify the value of the health and safety benefits that would arise from mandating the use of treated softwoods as such a task relies on assumptions being made about the incidence of injury or harm attributable to Borer damaged timbers. However, the modelling undertaken by the Department of Agriculture indicates that the introduction of statewide building restrictions could decrease the number of infested houses from around 198,000 (over the course of 30 years) to 77 (The Allen Consulting Group, 2006). Thus, the regulations could be seen to lower the risks of health and safety incidents occurring by a substantial degree.

(5) *Peace of mind*

The treatment of timbers to a level that would prevent EHB infestations may be a source of reassurance against attacks from other IAS. For instance, the treatments recommended for EHB also provide protection against termite infestations. This criterion captures the positive externality from this additional protection.

(6) *Damage to iconic structures*

The structural quality and longevity of houses does pose significant public interest concerns, particularly as houses are significant cultural or personal assets. Modern churches, public buildings and works of art and other structures of social significance may also be put at risk by EHB infestation.

Table 7.4.   Qualitative assessment results for the *Social* Criterion

	Management alternative		
Sub-criterion	Do nothing	Statewide building restrictions	Delimited building restrictions
Reduction in Infested Houses	VP	VG	VG
Peace of Mind	P	VG	G
Damage to Iconic Structures	VG	VG	VG
Unknown Catastrophe	VP	G	G

(7) *Unknown catastrophe*

This sub-criterion captures systemic risks brought about forces yet to be determined. The growing complexity of regulatory structures, corporations and institutions may make social systems less resilient to the impacts of a slow-spreading IAS like EHB in ways we are not yet aware of, but which could impose a significant cost on future generations.

Table 7.4 shows the jury's score for each management alternative using the scale defined in Table 7.1.

Environmental criteria

(8) *On and off-site damage* This criterion relates to chemical residue issues brought about through the increased use of insecticidal timber treatments. Chromated Copper-Arsenate (CCA) in particular leaches out of the treated timber over time so there can be residues of arsenic, copper and chromium on the surfaces of the wood. Timber treatment plants can be particularly contaminated (i.e. referred to by the jury as *on-site* contamination), but are covered by other environmental and chemical regulations. Nevertheless, an audit undertaken by the State of New South Wales Environment Protection Authority of five timber treatment plants found contamination through inadequate storage of materials and wastes at 5 plants, failure to maintain drains, dams

or treatment facility at 4 plants, and inadequate surface water controls at 4 plants (New South Wales Environment Protection Authority, 2003).

The European Commission Scientific Committee on Toxicity, Ecotoxicity and the Environment noted: 'There is extensive documentation of past substantial soil and groundwater contamination at wood treatment sites ... There is also evidence in the published literature ... that contamination of the soil and vegetation can extend to the area beyond the immediate boundaries of such sites, something that has been attributed to wind erosion, percolation, surface drainage as well as on-site incineration of wood waste' (Scientific Committee on Toxicity, 1998). Residues of arsenic, copper and chromium on the surfaces of the wood can be washed off by rain to accumulate beyond the confines of timber yards (i.e. referred to as *off-site* contamination of soil or groundwater). All three metals pose a potential threat to the environment. According to the US EPA: 'The amount and rate at which arsenic leaches, however, varies considerably depending on numerous factors, such as local climate, acidity of rain and soil, age of the wood product, and how much CCA was applied.' (Office of Pesticide Programs, 2002).

Table 7.5 shows the stakeholder jury's score for each management alternative using the scale defined in Table 7.1.

Three rounds of sub-criteria weighting were carried out in the workshop to allow for constructive debate about the relative importance of each to different stakeholders within the group. This process involved participants clearly articulating their preferences relative to others. The sub-criteria weights given by the jury in each

Table 7.5. Quantitative assessment results for the *Environmental* Criterion

Sub-criterion	Management alternative		
	Do nothing	Statewide building restrictions	Delimited building restrictions
On and off-site damage	VG	F	G

round are expressed using the terms described in Table 7.1. The linguistic results are given in Table 7.6. Note that these weights were assigned by the jury without consideration of the range of impact scores. Hence, in instances where each regulatory option was assigned the same score for a single criterion, the weight of that criterion becomes irrelevant. In a more comprehensive deliberative MCDA exercise the criterion would in fact be removed from the assessment.

With the data contained in Tables 7.3–7.5 we can determine the crisp performance index (i.e. P_i from Equation (7.12)) for each management alternative. Using a scale between 0 and 1, the cells of Table 7.6 indicate the Hamming distance for each management alternative. This reflects the performance of each option relative to the positive and negative ideal solutions for each criterion. The values indicated in each of the cells result from the summation of Equations (7.10) and (7.11). A crisp performance index and rank is shown for each alternative in the final two rows of Table 7.6. Note that the performance index (derived from Equation (7.12)) is shown as a percentage.

It is apparent from Table 7.7 (below) that the total performance index and the ranking of management alternatives changed very little across the three rounds of weighting. However, the process of deliberation between rounds one and two appears to have been the most significant in defining the preferred course of management action. Although equally as desirable to the *Statewide building restrictions* alternative in round one, preferences for the *do nothing* alternative were revised downwards in round two and remained low in round three. As indicated in Table 7.6, despite alterations to sub-criteria weightings between rounds two and three, these were not sufficient to produce a change in the ranking of management options.

Figure 7.1 shows the extent of weightings changes by round across the sub-criteria. These are expressed in percentage form, and individual criteria are grouped together along the horizontal axis. The sum of all the lightly shaded bars labelled 1 (i.e. round 1) equals 100%, as does the summation of all the bars labelled 2 and 3.

The individual members of the jury demonstrated a propensity to change their weighting opinions in response to information revealed

Table 7.6. Weighting of the sub-criteria by members of the stakeholder jury

		Economic sub-criteria			Social sub-criteria			Environmental sub-criteria	
	Jury member	Compliance costs	Expected damage costs	Administration costs	Reduction in infested houses	Peace of mind	Damage to iconic structures	Unknown catastrophe	On and off-site impact
Round 1	1	More	Most	Less	More	Less	Least	More	Fair
	2	More	Most	Fair	More	Most	Fair	More	Most
	3	More	Most	Least	Most	More	Least	Most	Least
	4	Fair	More	Less	More	More	Less	More	More
	5	Least	Most	Least	Most	More	Less	More	Least
	6	Less	More	Less	More	More	Fair	More	Fair
	7	Fair	More	Less	More	Fair	Fair	Most	Less
	8	More	More	Less	Fair	Most	Less	Most	Less
	9	Fair	More	Less	Less	More	Less	Most	Less
Round 2	1	Fair	Most	Least	Most	Fair	Least	Fair	Less
	2	More	Most	Fair	Fair	Most	Less	Fair	Fair
	3	More	Most	Least	Most	More	Least	Most	Least
	4	Fair	More	Fair	More	Most	Least	More	More
	5	Less	Fair	More	More	More	Least	Fair	Fair
	6	Fair	More	Less	More	More	Less	More	More
	7	Fair	Most	Less	More	More	Least	Most	Less
	8	More	More	Fair	More	More	Least	More	More
	9	Fair	More	Less	Fair	More	Less	Most	Less
Round 3	1	Fair	Most	Least	Less	Fair	Least	Less	Less
	2	More	Most	Fair	Fair	More	Least	Least	Fair
	3	Fair	Most	Less	Most	Fair	Least	Fair	Least
	4	More	Fair	Fair	Fair	More	Less	Fair	Fair
	5	More	More	More	More	More	Least	More	More
	6	Fair	More	Less	More	More	Less	More	More
	7	Fair	More	Less	More	More	Least	Fair	Least
	8	More	More	Fair	Fair	More	Least	Less	Less
	9	Less	More	Least	Fair	More	Least	Most	Less

Table 7.7. Total performance index

		Round 1			Round 2			Round 3		
		Do nothing	Statewide building restrictions	Delimited building restrictions	Do nothing	Statewide building restrictions	Delimited building restrictions	Do nothing	Statewide building restrictions	Delimited building restrictions
Economic sub-criteria	Compliance Costs	1.0	0.0	0.5	1.0	0.0	0.5	1.0	0.0	0.5
	Expected Damage Costs	0.0	1.0	1.0	0.0	1.0	1.0	0.0	1.0	1.0
	Administration Costs	1.0	0.0	1.0	1.0	0.0	0.0	1.0	0.0	0.0
Social sub-criteria	Reduction in Infested Houses	0.0	1.0	0.8	0.0	1.0	1.0	0.0	1.0	1.0
	Peace of Mind	0.0	1.0	1.0	0.0	1.0	0.8	0.0	1.0	0.8
	Damage to Iconic Structures	0.0	1.0	0.2	1.0	1.0	1.0	1.0	1.0	1.0
	Unknown Catastrophe	1.0	1.0	1.0	0.0	1.0	1.0	0.0	1.0	1.0
Environmental sub-criteria	On and Off-Site Impact	1.0	0.0	0.4	1.0	0.0	0.7	1.0	0.0	0.7
Total performance Index		30%	30%	40%	28%	33%	39%	28%	33%	39%
Rank		2	2	1	3	2	1	3	2	1

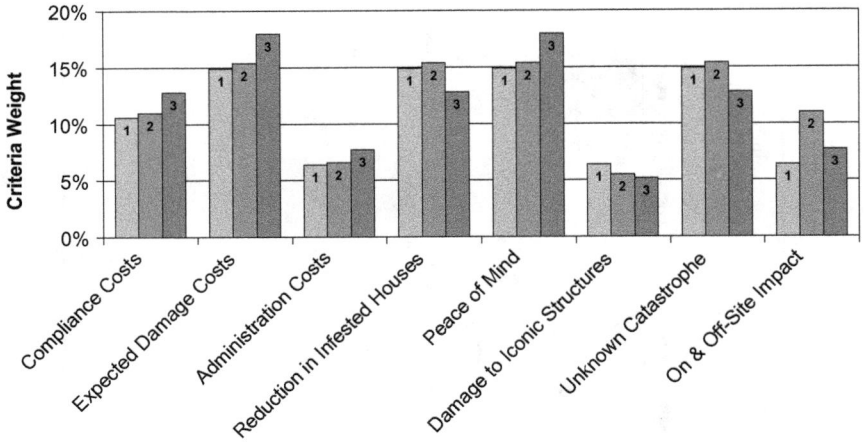

Figure 7.1. Change in sub-criteria weights by round

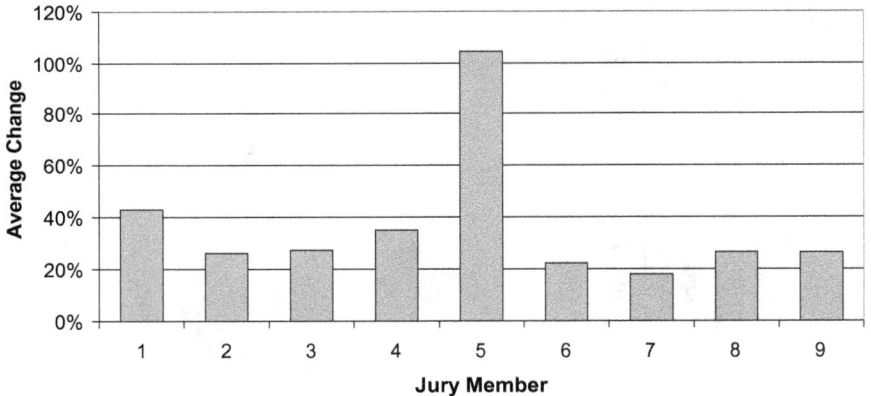

Figure 7.2. Change in sub-criteria weights by juror between rounds 1 and 3

and discussed during the deliberative sessions of the workshop. This is particularly the case with jury member 5, and to lesser extent members 1 and 4. But across the board weights were changed significantly over the course of the three rounds. Figure 7.2 illustrates the extent of weight changes for sub-criteria by jury member between rounds 1 and 3. Each bar indicates the average difference between an individual's set of weights in rounds 1 and 3 in absolute value terms.

7.3.4 *Issues with MCDA*

7.3.4.1 *Treatment of uncertainty*

The group interaction of deliberative MCDA makes it more challenging to incorporate uncertainties into decision making. The keys to success are: (a) to explicitly take uncertainty into consideration in the deliberative MCDA process, and (b) to eliminate linguistic uncertainty and other sources of arbitrary disagreement. In addition to the standard sensitivity analysis, there are two major ways of treating uncertainty explicitly in the MCDA literature. The first is to apply the fuzzy logic model (as above), and the second involves the inclusion of uncertainty-related criteria in forming the impact matrix (Cook and Proctor, 2007; White *et al.*, 2008; Kiker *et al.*, 2008).

When initially designing the deliberative MCDA case study, it was felt that including uncertainty-related criteria separately could be viewed as double counting or redundant. However, during the workshop the criterion *unknown catastrophe* was added by the jury to address the possibility of traumatic yet unseen impacts of EHB on future generations that are not addressed by the existing scientific research. The fuzzy system approach was adopted in an attempt to deal with the uncertainties associated with estimating values in the impact matrix, including model uncertainty, natural variation, measurement error and systematic error in an "all-in-one" manner. This is a coarse approach but it is difficult to separate these components when using other researchers' results in the impact matrix. The same fuzzy system approach was applied to handle the uncertainty due to the subjective judgement of workshop participants in ranking the criteria.

A five-tiered linguistic system (i.e. Tables 7.1 and 7.2) was used in an effort to reduce stakeholders' cognitive burden in the evaluation process, but we must point out several shortcomings of the scoring system used. Specifically, scoring alternatives on a generic 1-9 scale risks information that may be critical to a decision being discarded. It is often the case that jury members have a perfectly good understanding of natural units (e.g. $) that help them to capture this information in their scoring and weighting activities.

Moreover, replacing natural units with an assumed scale between the best and worst scores can introduce a scoring bias since it is not obvious what scale should be used. For instance, if a linear scale is used when a log scale would have been appropriate, the relative preference between different scores will be severely understated.

7.3.4.2 *The deliberative process*

We believe the deliberative MCDA method offers at least two advantages for biosecurity decision-support analysts. Firstly, it bridges the gap between risk assessment and risk management by encouraging a partnership between scientists and policymakers, or stakeholders. Secondly, it offers an opportunity to explore social dimensions of biosecurity risk before a collective decision is made.

The deliberative workshop process allows all outliers (i.e. those participants with the highest or lowest preferences) to state their rationale. This open discussion helps to address linguistic uncertainty. For instance, in our specific case study the criterion *unknown catastrophe* was renamed *long-term impacts* and *intergenerational impacts* during previous rounds of weighting to clarify what the criteria referred to. Deliberation also encourages group learning and information flows between scientists and stakeholders. For example, jury members learned that current chemicals used for timber treatments are much more environmentally friendly than in the past. As a result a drop in the weight for this criterion was observed after the deliberation process.

7.3.5 *Conclusions regarding deliberative MCDA from the case study*

By no means do we advocate the use of deliberative MCDA in every biosecurity policy decision. The EHB case study reveals at least three challenges in applying the technique to decide appropriate response actions following an incursion.

Firstly, deliberative processes pose new challenges in understanding how uncertainty affects decisions. It is difficult to understand why a group's preference changes, whether it is due to new exposure to uncertainty and different formats of presenting uncertain information, or due to other factors such as group composition and dynamics.

Secondly, the choice of a jury will no doubt directly affect decision outcomes (Cook and Proctor, 2007). As such, some argue that information based on a DMCE or participatory processes should not be used as the only source of preference information because it will inevitably represent the voice of the more active and opinionated population (Lahdelma *et al.*, 2000).

Thirdly, policymakers are generally unfamiliar with the deliberative process. As a result they may encounter difficulty in participating. For example, a compulsion to assign non-extreme weights to avoid being an outlier and having to justify or defend their opinions clearly biases preferences towards the 'middle ground'. To avoid this, there is a need for the facilitator to continuously emphasise that the purpose of the deliberative MCDA is not to generate uniform opinion. This emphasises the importance of the facilitator's expertise in conducting discussions to the success or otherwise of the process.

7.4 Conclusion

In this chapter, we have discussed some of the problems encountered when using cost–benefit analysis to guide IAS policy decisions. Specifically, we discussed problems related to coverage and consistency, uncertainty and valuation. To address these problems, or at least to make them transparent to policymakers, we put forward deliberative MCDA as a promising model for managing risks in the face of complex social values and profound uncertainty. The case study presented, showcased the use of deliberative MCDA with a fuzzy set approach to address an IAS response decision, but the same technique can be used in other multi-faceted risk management decision-making contexts, particularly when those risks have low probability, high novelty and high impacts (e.g. flood, earthquake, infectious diseases and abandoned hazardous waste dumps). However, it is important to stress that deliberative DMCE is not a replacement for conventional cost–benefit analysis, but is instead a means of reinforcing decisions and airing stakeholder preferences.

CHAPTER 8

SO FAR, SO GOOD...SO WHAT?

8.1 Introduction

A lot of ground has been covered in this book, and we hope to have provided some ideas as to how decision-support analysts can inform IAS policymakers in the absence of data and in the face of uncertainty. We have presented a generic ecological model that can be applied to a broad range of IAS and combined it with cost and revenue information to predict likely IAS costs and benefits of management over time. Based on this flexible model, we have presented a series of case studies and looked at the differences between the various taxonomic groups. We have also discussed methods that can be used to inform policies when the impacts of IAS extend to non-market goods such as social and environmental goods.

So what? How does what we have presented in this book change anything? Well, of course it does not, necessarily. We can build and refine the bioeconomic models, cost–benefit analyses and deliberative MCDA methods we use to our heart's content, but unless they do actually lead to better biosecurity decisions we will be in the same predicament we were to start with. And even that depends on what is meant by the descriptor, "better". There is a lot of linguistic uncertainty in such a term, after all.

In this final chapter, let us summarise what we think the implications and contributions of the book are from a broad policy perspective, hinting at what we think better biosecurity policymaking entails (without defining it specifically, in true fuzzy set style). We will discuss which needs of policymakers and their decision-support analysts we hope to have informed, and why our blend of bioeconomic modelling and interactive decision-facilitation methods might be of interest to them. We will also speculate as to what the future needs of technical decision-support specialists might be.

8.2 How Can Decision Makers Use This Book to Facilitate IAS Policy Decisions?

This book contains examples of probabilistic predictive models and decision-facilitation methods that at least begin to open up the complicated world of invasion biology and biosecurity economics to decision makers. In spite of the complexity, all-too-often policy decisions need to be made quickly in the wake of significant events, political pressure and the turbulent atmosphere of IAS incursions. We have been mindful of this when putting forward ideas that decision-support specialists might use to assist them in their difficult role; for changes resulting from policy decisions they inform today could affect generations to come in ways that are difficult to predict, and are probably irreversible.

Public officials and community stakeholders charged with the responsibility of making these decisions are often naive about what science can and cannot say about complex systems. In these situations, policymakers tend to rely on a limited number of "heuristic principles" to help them simplify the process of judgment (Kahneman and Knetsch, 1992). This means that without the help of analytical tools like those presented in this book, IAS policymaking will invariably suffer from problems such as omitting important criteria and fixing opinions based on insufficient information.

The impact simulation model showcased in Chapters 3 and 4 is put forward as a general tool that can be applied to a broad range of IAS, in contrast to more specific models for individual species. To date, the losses IAS of plants cause and the relationship between their abundance and costs, has been poorly specified (Ruiz *et al.*, 1999;

Gurevitch and Padilla, 2004; Thomas and Reid, 2007). We have made no attempt to collect historical data of IAS spread and impact around the world to rectify this problem, but have instead turned to predictive models to generate future scenarios.

Applying a well-known biological spread model to predict the behaviour of different species over time with and without controls, our results suggest total costs of many IAS affecting plants accrue over time in a linear fashion. That is, the costs associated with them tend to accrue at a constant rate over time. However, in certain cases in which affected hosts include large export-focused industries, a concave cost curve is produced over time. Here, costs tend to spike early in the invasion process and are eroded over time by a positive rate of discounting.

The situation is more complex with IAS that have environmental and social costs associated with them as well as agricultural costs, and are not well represented using our bioeconomic model on its own. While agricultural IAS generate immediate tangible losses, environmental IAS may spread extensively without having substantial impacts that can be valued. This means that there is likely to be a lag between achieving a level of environmental IAS abundance and realising the full losses this imposes on society. With these types of pests, we require a broader set of criteria on which to base policy than simply looking at agricultural costs over time.

The various agricultural IAS case studies put forward in this book clearly demonstrate the capacity of the bioeconomic model to generate useful and comparable impact predictions over significant time periods for a wide range of IAS. While the detailed parameterisation of these case studies has been carefully carried out, the information from which parameter estimates are formed is constantly changing as our knowledge of IAS and host environments increases. The case study information should therefore be seen as a snapshot reflecting the current state of knowledge. The information outputted by the bioeconomic model can evolve over time as our information base improves.

By integrating analytical tools and risk communication, deliberative MCDA can enhance the policy relevance of IAS cost estimates. Deliberative MCDA relies on gathering more diverse and context-specific knowledge from experts and stakeholders, exposing and debating the conditional social assumptions embedded within the models, and providing an opportunity to proactively prepare the ground for policy change (Penning-Rowsell *et al.*, 2006). This should mean that a decision based on integrated processes like those put forward in this book will gain more public trust and credibility than those made using more informal methods (Fischhoff, 1995).

Having said that, there are a number of challenging issues to address when applying the deliberative multi-criteria decision analysis (MCDA) in decision-facilitation for IAS risk. These include how a jury should be chosen, which can directly affect decision outcomes (Cook and Proctor, 2007). It may be argued that information based on a deliberative MCDA should not be used as the only source of preference information because it will inevitably represent the voice of more active and opinionated jury members. In addition, a jury member unfamiliar with the deliberative process may encounter difficulty in participating and interacting with experts (Renn, 2003), while a jury member familiar with the process may be prone to strategic misrepresentation of preferences. As in the case of valuation exercises of environmental economics, the deliberative MCDA process is also subject to the perils of information bias and "groupthink" (Ajzen *et al.*, 1996; Janis, 1982).

We do not advocate replacing traditional policy-support tools such as pest risk assessment and cost–benefit analysis with deliberative MCDA. On the contrary, we believe these tools can be integrated into the deliberative MCDA framework. For example, a cost–benefit-ratio or Net Present Value (NPV) may be used as one of the criteria regarding the desirability of different policy options for IAS management. We do argue that technical tools, as powerful as they can be, cannot completely solve environmental problems. This is because environmental decisions are political as well as scientific and resolving environmental problems requires addressing the values of the public (Beierle, 2002; Sarewitz, 2004). We believe this is particularly true when there is profound uncertainty in our scientific understanding.

Under this new model of deliberative MCDA-facilitated IAS risk management, scholars of biological invasion and risk analysts can integrate their research results into the decision-making process. They can provide expert testimony that communicates not only their research but also the uncertainty associated with their results to a policymaking jury. Essentially, this new decision-making model fits into a more democratic paradigm that conceptualises scientists as part of society, working with others to solve problems together (Norton, 1998; Larson, 2007; Robertson and Hull, 2003; Pielke, 2007). At the same time, the deliberative MCDA offers scientists an interactive platform where their work can be critically discussed and clearly articulated to policymakers.

8.3 How Do Our Methods Help Policymakers Choose Between Eradication and Prevention Activities?

As we mentioned at the outset of the book, IAS can be thought of as negative side effects of international trade. These side effects, or externalities, are created when the actions of some individuals and businesses influence the utility of other community members by exposing them to risk. Generally, the potential costs of these externalities are not reflected in the price of traded commodities. While the costs of an invasion are borne by parties outside the market for the 'risky' imported goods, consumers pay only the market clearing price, reflecting production costs and the profit margin of the supplier, plus transport to market.

The costs of an invasion may be felt in terms of market goods such as agricultural commodities, or non-market goods such as environmental and social goods. This means the unintended victims of IAS can be small in number, such as the members of an infant agricultural industry, or include every member of a community. Unlike consumers who have a choice of whether or not to consume an imported product, those subject to these wider impacts have little or no control over their exposure to incursion risk.

While aggregate regional and worldwide damage assessments have been useful in highlighting the severity of the IAS issue, quantitative information on specific IAS introductions and their impacts have not been widely available. This makes it difficult for policymakers to determine appropriate market failure corrective measures. The bioeconomic model developed and applied in Chapters 3–5 of this book represents a practical attempt to fill this gap, or to at least begin to quantify the size of the externality problems encountered with some IAS.

In Chapter 6 we discussed how the bioeconomic model can be used in cost–benefit analyses to calculate the expected benefits of prevention compared with control (chiefly eradication) responses to IAS incursions. We demonstrated how the costs and benefits of these response policy alternatives can be evaluated on the basis of the NPV they are expected to generate for society over time. However, in estimating NPV with a positive discount rate, we also showed that the length of time being considered is important in establishing priorities in which to invest our scarce biosecurity resources. Given this specification issue, the policy debate over prevention vs. eradication is set to continue!

We also saw that an important distinguishing feature of some IAS incursions is a loss of export market access, which can only be restored through eradication. The implications of this were explored in Section 6.4 and it was shown that in such cases there is the potential for ambiguity over whether a slowing-the-spread policy or eradication generates the greatest NPV of benefits. In particular, since eradication will always have greater response costs over the period eradication is achieved, a sub-eradication level of biosecurity investment may be preferred for eradication despite the loss of export markets. Nevertheless, it was also shown that if either the marginal benefits or the marginal costs of IAS removal are low up to the point of eradication, then the eradication option is more likely to be preferred. It was, therefore, argued that eradication will typically be preferred to control in situations where the export benefits of pest or disease-free status are the over-whelming consideration.

In planning risk management strategies for IAS that affect non-agricultural hosts and inflict environmental and social harm, a broader set of criteria are needed to decide an appropriate course of action. In Chapter 7 we demonstrated how a deliberative MCDA approach can be used in these sorts of cases using the example of the European house borer in WA. This combined indicators of ecological and socio-economic impact, and involved a jury of decision makers weighting criteria to indirectly rank response alternatives. By discussing each other's preferences and entering into dialogue to help them understand preferences within the jury, deliberative MCDA facilitates decisions that the collective jury is most comfortable with given the circumstances of the decision. While it does not replace conventional cost–benefit analysis, deliberative MCDA is a useful supplement for it in cases of environmental and social IAS.

In relation to policy implications, it is clear that in designing its response to an IAS outbreak policymakers need to consider each case on its own merits, but this is not necessarily consistent with current trends in IAS response management, as can be seen in the case of Australian biosecurity policy. Even when the impacts of IAS are market-based, applying cost–benefit analysis in Chapter 5 demonstrated the range of benefits that can be generated by different response strategies. However, Australian State-based biosecurity is being increasingly influenced by generalised, top–down approaches to decision making.

For example, one of the key instruments used currently in Australian IAS response decision making is the Generalised Invasion Curve, shown in Figure 8.1 (Agriculture Victoria, 2015). This stylised representation of IAS spreading over time suggests management strategies to employ in different IAS situations based purely on the area occupied, rather than any cost and benefit information. It is now used by State governments in WA (Department of Agriculture and Food Western Australia, 2015), New South Wales (NSW Department of Primary Industries, 2015) and Victoria (Agriculture Victoria, 2015) to guide investment in IAS risk management decisions.

The generalised invasion curve is an abstract sigmoid curve implying certainty in the relationship between the area affected by

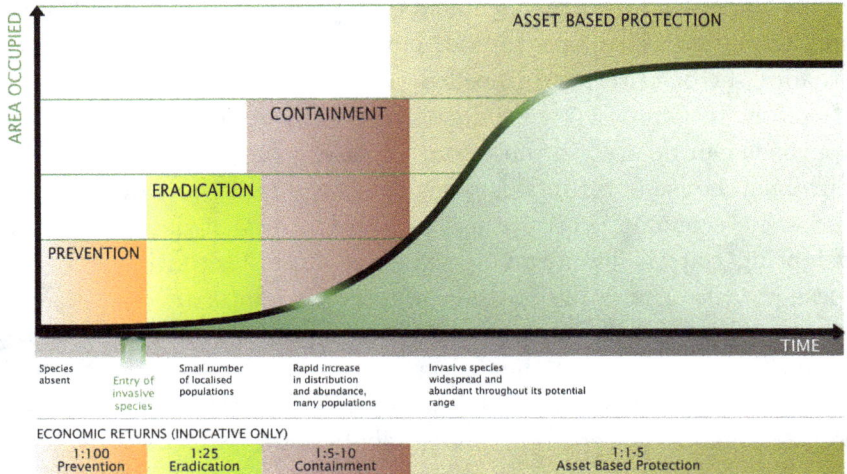

Figure 8.1. The Generalised Invasion Curve (Agriculture Victoria, 2015)

an IAS and the time since its arrival in a new region. Hypothetical cost-benefit ratios are offered at the bottom of the figure that suggest returns to investment decline with the area occupied, and that an appropriate intervention (particularly by governments) lies in prevention and eradication where returns are highest.

This attempt to simplify biosecurity policymaking does not consider uncertainty about the invasion process or cost and benefit information. For instance, if area freedom has a market access element to it and there is a spike in the total response benefit curve (as discussed in Chapter 6), we need not necessarily be on the region of the generalised invasion curve where eradication is indicated for this to be the preferred response.

Moreover, because externalities characterise all four areas of policy mentioned in Figure 8.1 (i.e. prevention, eradication, containment and asset-based protection), a case can be made for policy intervention in all of them, rather than just prevention and eradication as suggested by the indicative cost–benefit ratios. Indeed, if transaction costs associated with forming control targets and partnerships increase with area occupied, a rule-of-thumb for policy intervention from governments in particular may be the reverse of that suggested in the generalised invasion curve.

The very nature of complex systems means that they respond to shocks like IAS incursions in ways that are hard to understand and predict. Instead of an abstract screening device like the generalised invasion curve, the ideas put forward in this book advocate a case-by-case assessment of policy decisions. We do not take the view that bioeconomic models, cost–benefit analysis or deliberative MCDA are too expensive and time consuming for decision-support staff to use. Rather, the generic approaches we have demonstrated are intended to make these decision aids more accessible and useful to time-pressured policymakers. In particular, their ability to communicate uncertainty to decision makers and the implications of policies over time we feel will produce more informed policies than those simply based on guides like the generalised invasion curve.

8.4 How Might Decision Support Tools and Methods Change in the Future?

Social media now plays a sizeable role in decision making regarding IAS, and makes both a positive and a negative contribution. Rapid advancement in social media sites and handheld information technologies has hugely increased the dissemination of information, but unfortunately the quality of the information can be hard to determine. Anecdotal, unsubstantiated information is difficult to expose as such before it reaches decision makers and begins to affect resource allocation decisions. This problem is compounded by the quantity of information reaching decision makers via their information networks and the apparent decreasing benefits to intended users.

When an event occurs in the "here and now" decision makers of today have greater access to information about it than ever before, but as a result think of it in concrete, low-level (i.e. intricate detail) terms. When an event like an IAS incursion is further removed from direct experience (i.e. is more distant into the future), decision makers have less available and reliable information about it and form more abstract and schematic representations of the event (Trope *et al.*, 2007).

In terms of the type of information of most use to decision makers, pictures (as opposed to words and statistics) are concrete representations that carry the properties of an invasion event in full detail (Liberman and Trope, 1998; Liberman *et al.*, 2002). This seems to imply that when a decision-making group is psychologically 'near' to an event, pictorial representations of it may be more effective decision aids than words and statistics (Förster *et al.*, 2004).

So, perhaps decision-support analysts of the future will begin to explore more deeply information networks and adopt a learning-by-doing approach to policy formation. Bioeconomic models like those put forward in this book can be useful in communicating complex phenomena to decision makers, but tend to be limited to non-spatial decision contexts due to a lack of specificity. Where host environments are homogenous this does not really pose a problem, but in agriculturally and environmentally diverse regions the spatial characteristics of IAS impacts can be highly varied, and important in a policymaking context.

Future decision support models might be employed to simulate intertemporal effects of IAS across different landscapes in real time using maps to communicate information, in addition to traditional statistical indicators. This would allow exercises to be conducted placing decision makers in virtual worlds that they must protect against IAS. By supplying them with information about IAS and the context in which they are to mount a response against an invasion, decision makers could potentially hone skills they might need to manage real incursion events.

If so, interactive, map-based models could change the way response policies are developed (Harwood *et al.*, 2009; Liu *et al.*,

2015; Cook *et al.*, 2016). Potentially, they could replace reactionary, backward-looking policymaking with an adaptive approach. To do so, of course, they would need to feed into a management framework that thinks about IAS as part of the dynamics of larger social-ecological systems, and learns through policy trial and error in virtual worlds where mistakes are costless. Policymakers themselves could then approach real world IAS problems proactively, with a more complete understanding of how their decisions are likely to affect the broader system.

8.5 Conclusion

We hope you have found this book to be of interest, and have come away with a host of thoughts and ideas about the IAS management problem. The diversity of IAS threats facing agricultural industries around the world makes planning for future biosecurity scenarios extremely difficult. However, since we are dealing with agricultural and ecological systems that are complex, future prediction is fraught with difficulty.

Decision makers who face demands from different government and private institutions for biosecurity support have a difficult time determining the relative importance of investment options. This is even true at a species level despite the common IAS characteristics of introduction, spread and impact that determine the strategic importance of managing their effects. The approaches developed in this book provide decision makers, be they private or government, with a means to undertake quantitative and qualitative policy comparisons. Amongst other things, it has included: a theoretical discussion of quantitative impact simulation modelling; a technical discussion of a quantitative bio-economic model to calculate IAS impacts on plant-based industries over time; case studies to demonstrate the model's breadth of application; a critique of the cost–benefit analysis approach to policy evaluation and prioritisation; and a group-based deliberative approach to decision-making involving non-market IAS impacts.

From our discussions, it is clear that there is a great deal of uncertainty about IAS risks facing us now and in the future. With the continued growth of international trade, cargo and people movements and unregulated trade, IAS risks facing any agricultural region will almost certainly increase over time, necessitating more and more resources to cope. However, predictions about those risks and the costs of mitigation measures can be made using a bioeconomic impact simulation model like that presented in this book. This enables us to track likely impacts over time and to compare them with those of other species so that they can be prioritised in terms of their potential cost to society. When the impacts of IAS are limited to market-based effects, conventional cost–benefit analysis can be used in conjunction with these models to inform policymakers as to the best risk-mitigation strategies to follow. When non-market impacts are a factor, they can be considered alongside market effects using MCDA methods to prioritise policies. Where diverse mixes of stakeholders are (potentially) affected by an IAS, deliberative processes can be used.

There is no question that IAS and biosecurity generally are complicated areas of study. There is a wealth of skills and ideas that can be drawn in to shed new light on issues of risk management, data analysis and response strategies, but without an ability to interpret and understand this information it is at risk of being overlooked. By building sufficient flexibility into tools such as those presented in this book, we hope to make a contribution that benefits society in terms of future biosecurity decision making.

REFERENCES

AUSTRALIAN BUREAU OF AGRICULTURAL AND RESOURCE ECO-
NOMICS AND SCIENCE. (ABARES) 2011. *Agricultural Commodity
Statistics 2011*. Canberra: Australian Bureau of Agricultural and Resource
Economics and Sciences.

AUSTRALIAN BUREAU OF STATISTICS (ABS) 2012. *Value of Agricultural
Commodities Produced, Australia, 2010–2011*. Canberra: Australian Bureau
of Statistics.

AUSTRALIAN BUREAU OF STATISTICS (ABS) 2014. Agricultural Commodities, Australia, 2011–2012. Canberra: Australian Bureau of Statistics.

ADAMOWICZ, W. L. 2004. What's it worth? An examination of historical trends
and future directions in environmental valuation. *The Australian Journal of
Agricultural and Resource Economics*, 48, 419–443.

AGASSIZ, D. J. L. 1994. *Population Dynamics of Invading Insects*. PhD, Imperial
College London.

AGRICULTURE VICTORIA 2015. *Invasive Plants and Animals Policy Framework*. Melbourne: Agriculture Victoria.

AHMEDANI, M. S., HAQUE, M. I., AFZAL, S. N., NAEEM, M., HUSSAIN, T. &
NAZ, S. 2011. Quantitative losses and physical damage caused to wheat
kernel (*Triticum aestivum* L.) by khapra beetle infestation. *Pakistan Journal
of Botany*, 43, 659–668.

AJZEN, I., BROWN, T. C. & ROSENTHAL, L. H. 1996. Information bias in
contingent valuation: Effects of personal relevance, quality of information,
and motivational orientation. *Journal of Environmental Economics and
Management*, 30, 43–57.

ALLAN, J., LOW CHOY, S. & MENGERSEN, K. 2010. Elicitator: An expert
elicitation tool for regression in ecology. *Environmental Modelling & Software*,
25, 129–145.

ANDERSEN, M. C., ADAMS, H., HOPE, B. & POWELL, M. 2004. Risk
assessment for invasive species. *Risk Analysis*, 24, 787–793.

ANDOW, D., KAREIVA, P., LEVIN, S. & OKUBO, A. 1993. *Spread of Invading
Organisms: Patterns of Spread*, New York: Wiley.

ANDOW, D. A., KAREIVA, P. M., LEVIN, S. & OKUBO, A. 1990. Spread of
invading organisms. *Landscape Ecology*, 4, 177–188.

AUSTRALIAN OILSEEDS FEDERATION (AOF) 2012. *Quality Standards, Technical Information and Typical Analysis 2012/13*. Sydney: Australian Oilseeds Federation Incorporated.

AUGUSTIN, S., GUICHARD, S., SVATOŠ, A. & GILBERT, M. 2004. Monitoring the regional spread of the invasive leafminer *Cameraria ohridella* (Lepidoptera: Gracillariidae) by damage assessment and pheromone trapping. *Environmental Entomology*, 33, 1584–1592.

AUSTRALIAN WEEDS COMMITTEE 2012. Weeds of National Significance: Rubber vine (*Cryptostegia grandiflora* Roxb. ex R.Br.) strategic plan 2012–17. Canberra: Australian Government Department of Agriculture, Fisheries and Forestry.

AYLOR, D. 2003. Spread of plant disease on a continental scale: Role of aerial dispersal of pathogens. *Ecology*, 84, 1989–1997.

BAKER, R. H. A., BLACK, R., COPP, G. H., HAYSOM, K. A., HULME, P. E., THOMAS, M. B., BROWN, A., BROWN, M., CANNON, R. J. C., ELLIS, J., ELLIS, M., FERRIS, R., GLAVES, P., GOZLAN, R. E., HOLT, J., HOWE, L., KNIGHT, J. D., MACLEOD, A., MOORE, N. P., MUMFORD, J. D., MURPHY, S. T., PARROTT, D., SANSFORD, C. E., SMITH, G. C., ST-HILAIRE, S. & WARD, N. L. 2008. The UK risk assessment scheme for all non-native species. Neobiota — From Ecology to Conservation: The 4th European Conference of the Working Group NEOBIOTA on Biological Invasions, 27–29 September 2006–2008 Vienna, Austria. 46–57.

BAKER, R. H. A., SANSFORD, C. E., JARVIS, C. H., CANNON, R. J. C., MACLEOD, A. & WALTERS, K. F. A. 2000. The role of climatic mapping in predicting the potential geographical distribution of non-indigenous pests under current and future climates. *Agriculture, Ecosystems & Environment*, 82, 57–71.

BANKS, H. J. 1977. Distribution and establishment of *Trogoderma granarium* Everts (Coleoptera: Dermestidae): Climatic and other influences. *Journal of Stored Products Research*, 13, 183–202.

BARRY, S., COOK, D., DUTHIE, R., CLIFFORD, D. & ANDERSON, D. 2010. *Future Surveillance Needs for Honeybee Biosecurity*, Canberra: Rural Industries Research and Development Corporation.

BASKIN, Y. 2002. *A Plague of Rats and Rubber Vines: The Growing Threat of Species Invasions*, London: Island Press.

BEIERLE, T. C. 2002. The quality of stakeholder-based decisions. *Risk Analysis*, 22, 739–749.

BELLMAN, R. E. & ZADEH, L. A. 1970. Decision-making in a fuzzy environment. *Management science*, 17, B-141–B-164.

BELTON, D. 2000. How the varroa management programme works. *Biosecurity*, 21, 5–6.

BENITEZ, J. M., MARTIN, J. C. & ROMAN, C. 2007. Using fuzzy number for measuring quality of service in the hotel industry. *Tourism Management*, 28, 544–555.

BERGSTROM, J. C. & DE CIVITA, P. 1999. Status of Benefits Transfer in the United States and Canada: A Review. *Canadian Journal of Agricultural Economics/Revue canadienne d'agroeconomie*, 47, 79–87.

BHATI, U. N. & REES, C. 1996. Fire blight: A cost analysis of importing apples from New Zealand. Canberra: Australian Bureau of Agricultural and Resource Economics.

BIOSECURITY AUSTRALIA 2001. Guidelines for import risk analysis. Canberra: Agriculture, Fisheries and Forestry Australia/Biosecurity Australia.

BIOSECURITY AUSTRALIA 2006. Final import risk analysis report for apples from New Zealand, Part A. Canberra: Biosecurity Australia.

BLAMEY, R., ROLFE, J., BENNETT, J. & MORRISON, M. 2000. Valuing remnant vegetation in Central Queensland using choice modelling. *Australian Journal of Agricultural and Resource Economics*, 44, 439–456.

BOMFORD, M. & HART, Q. 2002. Non-indigenous vertebrates in Australia. *In:* PIMENTEL, D. (ed.) *Biological Invasions: Economic and Environmental Costs of Alien Plant, Animal, and Microbe Species.* London: CRC Press.

BONDE, M. R., PETERSON, G. L., SCHAAD, N. W. & SMILANICK, J. L. 1997. Karnal bunt of wheat. *Plant Disease*, 81, 1370–1377.

BORN, W., RAUSCHMAYER, F. & BRÄUER, I. 2005. Economic evaluation of biological invasions — a survey. *Ecological Economics*, 55, 321–336.

BOSSENBROEK, J. M., MCNULTY, J. & KELLER, R. P. 2005. Can ecologists heat up the discussion on invasive species risk? *Risk Analysis*, 25, 1595–1597.

BOTHA, J., HARDIE, D. & CASELLA, F. 2004. Sawflies: The wheat stem sawfly *Cephus cinctus* and relatives: Exotic threat to Western Australia. GrainGuard Initiative. Factsheet No. 12/2004, South Perth: Department of Agriculture, Western Australia.

BRENNAN, J. P., WARHAM, E. J. & BYERLEE, D. 1992. Evaluating the economic impact of quality-reducing, seed-borne disease: Lessons from Karnal Bunt of wheat. *Agricultural Economics*, 6, 345–352.

BRADSHAW, G. A. & BORCHERS, J. G. 2000. Uncertainty as Information: Narrowing the Science-policy Gap. *Ecology and Society*, 4, 7. Available at: http://www.consecol.org/vol4/iss1/art7/.

BRIGHT, C. 1998. *Life out of Bounds: Bioinvasion in a Borderless World*, London: W.W. Norton & Co.

BRODEUR, J., LECLERC, L.-A., FOURNIER, M. & ROY, M. 2001. Cabbage seedpod weevil (Coleoptera: Curculionidae): New pest of canola in North-eastern North America. *The Canadian Entomologist*, 133, 709–711.

BURGMAN, M. 2005. *Risks and Decisions for Conservation and Environmental Management*, Cambridge: Cambridge University Press.

BURGMAN, M. A., KEITH, D. A. & WALSHE, T. V. 1999. Uncertainty in comparative risk analysis for threatened Australian plant species. *Risk Analysis*, 19, 585–598.

BUSCHMAN, L. L. 1976. Invasion of Florida by the "lovebug" *Plecia nearctica* (Diptera: Bibionidae). *Florida Entomologist*, 59, 191–194.

CAMPBELL, S. & WILKINSON, I. 2014. Business Case: Informing future investment in the management of starlings in Western Australia, 3rd Draft (4th July). South Perth: Department of Agriculture and Food, Western Australia.

CAREY, J. M. & BURGMAN, M. A. 2008. Linguistic uncertainty in qualitative risk analysis and how to minimize it. *Annals of the New York Academy of Sciences*, 1128, 13–17.

CENTRE FOR AGRICULTURAL BIOSCIENCE INTERNATIONAL 2014. Invasive Species Compendium. Cayman Islands: CAB International.

CHANG, Y.-H. & YEH, C.-H. 2002. A survey analysis of service quality for domestic airlines. *European Journal of Operational Research*, 139, 166–177.

CHAPMAN, T. 2005. The Status and Impact of the Rainbow Lorikeet (*Trichoglossus Haematodus Moluccanus*) in South-West Western Australia, *Miscellaneous Publication 04/2005*. South Perth: Department of Agriculture, Western Australia.

CHAPMAN, T. & MASSAM, M. 2007. Pestnote: Rainbow lorikeet. South Perth: Department of Agriculture and Food, Western Australia.

CLARK, J. S., CARPENTER, S. R., BARBER, M., COLLINS, S., DOBSON, A., FOLEY, J. A., LODGE, D. M., PASCUAL, M., PIELKE, R., PIZER, W., PRINGLE, C., REID, W. V., ROSE, K. A., SALA, O., SCHLESINGER, W. H., WALL, D. H. & WEAR, D. 2001. Ecological forecasts: An emerging imperative. *Science*, 293, 657–660.

COMMONWEALTH OF AUSTRALIA 2006. Handbook of Cost–Benefit Analysis. Canberra: Department of Finance and Administration.

CONVENTION ON BIOLOGICAL DIVERSITY 1991. *Convention Text* [Online]. Convention on Biological Diversity. Available at: http://www.biodiv.org/convention/articles.asp. [Accessed 20 December 2005].

COOK, D. C. 2003. Prioritising exotic pest threats to Western Australian plant industries. *Discussion Paper*. Bunbury: Government of Western Australia-Department of Agriculture.

COOK, D. C., AURAMBOUT, J.-P., VILLALTA, O. N., LIU, S., EDWARDS, J. & MAHARAJ, S. 2016. A bio-economic 'war game' model to simulate plant disease incursions and test response strategies at the landscape scale. *Food Security*, 8, 37–48.

COOK, D. C. & FRASER, R. W. 2008. Trade and invasive species risk mitigation: Reconciling WTO compliance with maximising the gains from trade. *Food Policy*, 33, 176–184.

COOK, D. C. & FRASER, R. W. 2014. Eradication versus control of Mediterranean fruit fly in Western Australia. *Agricultural and Forest Entomology*, 17, 173–180.

COOK, D. C., FRASER, R. W., PAINI, D. R., WARDEN, A. C., LONSDALE, W. M. & BARRO, P. J. D. 2011a. Biosecurity and yield improvement technologies are strategic complements in the fight against food insecurity. *PLoS ONE*, 6, e26084.

COOK, D. C., FRASER, R. W., WAAGE, J. K. & THOMAS, M. B. 2011b. Prioritising biosecurity investment between agricultural and environmental systems. *Journal of Consumer Protection and Food Safety*, 6, 3–13.

COOK, D. C., HURLEY, M., LIU, S., SIDDIQUE, A.-B. M., LOWELL, K. E. & DIGGLE, A. 2010. Final Report CRC10010 Enhanced Risk Analysis Tools. Canberra: Cooperative Research Centre for National Plant Biosecurity.

COOK, D. C. & PROCTOR, W. L. 2007. Assessing the threat of exotic plant pests. *Ecological Economics*, 63, 594–604.

COOK, D. C., THOMAS, M. B., CUNNINGHAM, S. A., ANDERSON, D. L. & DE BARRO, P. J. 2007. Predicting the economic impact of an invasive species on an ecosystem service. *Ecological Applications*, 17, 1832–1840.

COOK, D. R., CAMPBELL, G. W. & MELDRUM, A. R. 1990. Suspected *Cryptostegia grandiflora* (rubber vine) poisoning in horses. *Australian Veterinary Journal*, 67, 344.

COSTANZA, R., DE GROOT, R., SUTTON, P., VAN DER PLOEG, S., ANDERSON, S. J., KUBISZEWSKI, I., FARBER, S. & TURNER, R. K. 2014. Changes in the global value of ecosystem services. *Global Environmental Change*, 26, 152–158.

CRC FOR AUSTRALIAN WEED MANAGEMENT 2003. Weed Management Guide: Rubervine, *Cryptostegia grandiflora*. Canberra: Cooperative Research Centre for Australian Weed Management and the Commonwealth Department of the Environment and Heritage.

CROSBY, N. 1999. Using the citizens jury process for environmental decision making. *In:* SEXTON, K., MARCUS, A., EASTER, K. & BURKHARDT, T. (eds.) *Better Environmental Decisions: Strategies for Governments, Businesses and Communities.* Washington DC: Island Press.

CROWL, T. A., CRIST, T. O., PARMENTER, R. R., BELOVSKY, G. & LUGO, A. E. 2008. The spread of invasive species and infectious disease as drivers of ecosystem change. *Frontiers in Ecology and the Environment*, 6, 238–246.

CUNNINGHAM, R. J. 2012. Final Report CRC20137 Khapra beetle Diagnostics. Canberra: Cooperative Research Centre for National Plant Biosecurity.

CURTIS, K. & MCCORMICK, B. 2012. Western Australian Beef Commentary. Bunbury: Department of Agriculture and Food, Western Australia.

DAHLSTROM DAVIDSON, A., CAMPBELL, M. L. & HEWITT, C. L. 2013. The role of uncertainty and subjective influences on consequence assessment by aquatic biosecurity experts. *Journal of Environmental Management*, 127, 103–113.

DAVIDSON, M. D. 2013. On the relation between ecosystem services, intrinsic value, existence value and economic valuation. *Ecological Economics*, 95, 171–177.

DAWA 1997. *Farmnote 97/96: Bedstraw.* South Perth: Department of Agriculture Western Australia.

DE FRANÇA DORIA, M., BOYD, E., TOMPKINS, E. L. & ADGER, W. N. 2009. Using expert elicitation to define successful adaptation to climate change. *Environmental Science & Policy*, 12, 810–819.

DE MILLIANO, J. W., WOOLNOUGH, A., REEVES, A. & SHEPHERD, D. 2010. Ecologically significant invasive species: A monitoring framework for natural resource management groups in Western Australia, *Prepared for the Natural Heritage Trust 2 Program*. South Perth: Department of Agriculture and Food, Western Australia.

DEPARTMENT OF AGRICULTURE AND FOOD, WESTERN AUSTRALIA 2012. *An Analysis of Production Capability and Market Demand for Selected Western Australian Agricultural Products in 2012*. South Perth: Department of Agriculture and Food, Western Australia.

DEPARTMENT OF AGRICULTURE AND FOOD WESTERN AUSTRALIA 2015. *Invasive Species Plan for Western Australia 2015–2019*. South Perth: Department of Agriculture and Food Western Australia.

DIENEL, P. & RENN, O. 1995. Planning cells: A gate to fractal mediation. *In:* RENN, O., WEBLER, T. & WIEDEMANN, P. (eds.) *Fairness and Competition in Citizen Participation*. Dordrecht: Kluwer Academic Publishers.

DOAK, A., DEVEZE, M., MARCH, N., OSMOND, R. & MCKENZIE, J. 2004. Rubber vine management, *Control Methods and Case Studies*. Rockhampton: Department of Natural Resources, Mines and Energy.

DOBES, L. & BENNETT, J. 2009. Multi-Criteria Analysis: "Good Enough" for Government Work? *Agenda: A Journal of Policy Analysis and Reform*, 7–29.

DWYER, G. 1992. On the spatial spread of insect pathogens — theory and experiment. *Ecology*, 73, 479–494.

EDWARD, A. & KINGWELL, R. 2003. Bedstraw (*Galium tricornutum*) Eradication: Economic Analysis. South Perth: Department of Agriculture Western Australia.

EVANS, A. M. & GREGOIRE, T. G. 2007. A geographically variable model of hemlock woolly adelgid spread. *Biological Invasions*, 9, 369–382.

FARBER, D. A. 2003. Probabilities behaving badly: Complexity theory and environmental uncertainty. *Environs: Environmental Law and Policy Journal*, 27, 145–173.

FETENE, G. M., OLSEN, S. B. & BONNICHSEN, O. 2014. Disentangling the pure time effect from site and preference heterogeneity effects in benefit transfer: An empirical investigation of transferability. *Environmental and Resource Economics*, 59, 583–611.

FIELDING, N. J., EVANS, H. F., WILLIAMS, J. M. & EVANS, B. 1991. Distribution and spread of the great European spruce bark beetle, *Dendroctonus micans*, in Britain — 1982 to 1989. *Forestry*, 64, 345–358.

FINNOFF, D., SHOGREN, J. F., LEUNG, B. & LODGE, D. 2007. Take a risk: Preferring prevention over control of biological invaders. *Ecological Economics*, 62, 216–222.

FISHER, R. A. 1937. The wave of advance of advantageous genes. *Annual Eugenics*, 7, 353–369.

FÖRSTER, J., FRIEDMAN, R. S. & LIBERMAN, N. 2004. Temporal construal effects on abstract and concrete thinking: Consequences for insight and creative cognition. *Journal of Personality and Social Psychology*, 87, 177–189.

FRANKLIN, J., SISSON, S. A., BURGMAN, M. A. & MARTIN, J. K. 2008. Evaluating extreme risks in invasion ecology: Learning from banking compliance. *Diversity and Distributions*, 14, 581–591.

GATT 1994. Agreement on the Application of Sanitary and Phytosanitary Measures. *The Results of the Uruguay Round of Multilateral Trade Negotiations: The Legal Texts*. Geneva: General Agreement on Tariffs and Trade Secretariat.

GONG, W., SINDEN, J., BRAYSHER, M. & JONES, R. 2009. *The economic impacts of vertebrate pests in Australia*. Canberra: Invasive Animals Cooperative Research Centre.

GORDON, D. R., ONDERDONK, D. A., FOX, A. M. & STOCKER, R. K. 2008. Consistent accuracy of the Australian weed risk assessment system across varied geographies. *Diversity and Distributions*, 14, 234–242.

GROVES, R. H., BODEN, R. & LONSDALE, W. M. 2005. Jumping the Garden Fence: Invasive garden plants in Australia and their environmental and agricultural impacts, *WWF-Australia*. Sydney: CSIRO Report for WWF-Australia.

GROVES, R. H. & HOSKING, J. R. 1998. *Recent Incursions of Weeds to Australia 1971–1995*. CRC for Weed Management Systems Technical Series No. 3. Glen Osmond, S.A.: CRC for Weed Management System.

GUREVITCH, J. & PADILLA, D. K. 2004. Are invasive species a major cause of extinctions? *Trends in Ecology and Evolution*, 19, 470–474.

HAFI, A., REYNOLDS, R. & OLIVER, M. 1994. Economic Impact of Newcastle Disease on the Australian Poultry Industry, ABARE Research Report 94.7. A. B. O. A. A. R. Economics. Canberra: Australian Government Publishing Service.

HAJKOWICZ, S. & COLLINS, K. 2007. A review of multiple criteria analysis for water resource planning and management. *Water Resource Management*, 21, 1553–1566.

HANLEY, N., MACMILLAN, D., PATTERSON, I. & WRIGHT, R. E. 2003. Economics and the design of nature conservation policy: A case study of wild goose conservation in Scotland using choice experiments. *Animal Conservation*, 6, 123–129.

HARGROVE, J. W. 2000. A theoretical study of the invasion of cleared areas by tsetse flies (Diptera: Glossinidae). *Bulletin of Entomological Research*, 90, 201–209.

HARWOOD, T. D., XU, X., PAUTASSO, M., JEGER, M. J. & SHAW, M. W. 2009. Epidemiological risk assessment using linked network and grid based modelling: *Phytophthora ramorum* and *Phytophthora kernoviae* in the UK. *Ecological Modelling*, 220, 3353–3361.

HASTINGS, A., CUDDINGTON, K., DAVIES, K. F., DUGAW, C. J., ELMENDORF, S., FREESTONE, A., HARRISON, S., HOLLAND, M., LAMBRINOS, J., MALVADKAR, U., MELBOURNE, B. A., MOORE, K., TAYLOR, C. & THOMSON, D. 2005. The spatial spread of invasions: New developments in theory and evidence. *Ecology Letters*, 8, 91–101.

HENGEVELD, B. 1989a. *Dynamics of Biological Invasions*, London: Chapman and Hall.

HENGEVELD, R. 1989b. *Dynamics of Biological Invasions*, London: Chapman and Hall.

HINCHY, M. D. & FISHER, B. S. 1991. *A Cost–Benefit Analysis of Quarantine*, Canberra: Australian Bureau of Agricultural and Resource Economics.

HINCHY, M. D. & LOW, J. 1990. Cost–Benefit Analysis of Quarantine Regulations to Prevent the Introduction of Fire Blight into Australia: Report to the Australian Quarantine and Inspection Service. Canberra: Australian Bureau of Agricultural and Resource Economics.

HOLMES, E. E. 1993. Are diffusion-models too simple — a comparison with telegraph models of invasion. *American Naturalist*, 142, 779–795.

HOMAN, H. W. & MCCAFFREY, J. P. 1993. Insect pests of spring-planted canola. *Current Information Series (USA)*, 982.

HORAN, R. D., PERRINGS, C., LUPI, F. & BULTE, E. H. 2002. The economics of invasive species management: Biological pollution prevention strategies under ignorance: The Case of Invasive Species. *American Journal of Agricultural Economics*, 84, 1303–1310.

HORN, F. P. & BREEZE, R. G. 2000. Agriculture and food security. *In:* FRAZIER, T. W. & RICHARDSON, D. C. (eds.) *Food and Agricultural Security: Guarding Against Natural Threats and Terrorist Attacks Affecting Health, National Food Supplies, and Agricultural Economics*. 1st edn. New York: New York Academy of Sciences.

HUNTER, C., JOHNSON, K. & OSMOND, R. 2008. Rabbit control in Queensland: A guide for land managers. Brisbane: Department of Primary Industries and Fisheries.

JANIS, I. 1982. *Groupthink*, Boston: Houghton Mifflin.

JUANG, C. H. 1991 & LEE, D. H. 1991. A fuzzy scale for measuring weights of criteria in hierarchical structures. *In:* TERANO, T., (ed). *Fuzzy Engineering Toward Human Friendly Systems: Proceedings of the International Fuzzy Engineering Symposium*, Yokohama, Japan. IOS Press. 415–421.

KAHNEMAN, D. & KNETSCH, J. 1992. Valuing public goods: The purchase of moral satisfaction. *Journal of Environmental Economics and Management*, 22, 57–70.

KANGAS, A. S. & KANGAS, J. 2004. Probability, possibility and evidence: Approaches to consider risk and uncertainty in forestry decision analysis. *Forest Policy and Economics*, 6, 169–188.

KAREIVA, P. 1983. Local movement in herbivorous insects: Applying a passive diffusion model to mark-recapture field experiments. *Oecologia*, 57, 322–327.

KIKER, G. A., BRIDGES, T. S. & KIM, J. 2008. Integrating comparative risk assessment with multi-criteria decision analysis to manage contaminated sediments: An example for the New York/New Jersey harbor. *Human and Ecological Risk Assessment*, 14, 495–511.

KOLAR, C. S. & LODGE, D. M. 2002. Ecological predictions and risk assessment for alien fishes in North America. *Science*, 298, 1233–1236.

KUHNERT, P. M., MARTIN, T. G. & GRIFFITHS, S. P. 2010. A guide to eliciting and using expert knowledge in Bayesian ecological models. *Ecology Letters*, 13, 900–914.

LARSON, B. M. H. 2007. An alien approach to invasive species: Objectivity and society in invasion biology. *Biological Invasions*, 9, 947–956.

LEUNG, B., LODGE, D. M., FINNOFF, D., SHOGREN, J. F., LEWIS, M. A. & LAMBERTI, G. 2002. An ounce of prevention or a pound of cure: Bioeconomic risk analysis of invasive species. *Proceedings of the Royal Society B*, 269, 2407–2413.

LEVINE, J. M. & D'ANTONIO, C. 2003. Forecasting biological invasions with increasing international trade. *Conservation Biology*, 17, 322–326.

LEWIS, M. A. 1997. Variability, patchiness, and jump dispersal in the spread of an invading population. *In:* TILMAN, D. & KAREIVA, P. (eds.) *Spatial Ecology: The Role of Space in Population Dynamics and Interspecific Interactions*. New Jersey: Princeton University Press.

LIBERMAN, N., SAGRISTANO, M. & TROPE, Y. 2002. The effect of temporal distance on level of construal. *Journal of Experimental Social Psychology*, 38, 523–535.

LIBERMAN, N. & TROPE, Y. 1998. The role of feasibility and desirability considerations in near and distant future decisions: A test of temporal construal theory. *Journal of Personality and Social Psychology*, 75, 5–18.

LIEBHOLD, A. M. 2012. Forest pest management in a changing world. *International Journal of Pest Management*, 58, 289–295.

LIEBHOLD, A. M. & TOBIN, P. C. 2008. Population ecology of insect invasions and their management. *Annual Review of Entomology*, 53, 387–408.

LINE, R. F. 2002. Stripe rust of wheat and barley in North America: A retrospective historical review 1. *Annual Review of Phytopathology*, 40, 75–118.

LINKOV, I., VARGHESE, A., JAMIL, S., SEAGER, T. P., KIKER, G. & BRIDGES, T. 2004. Multi-Criteria Decision Analysis: A framework for structuring remedial decisions at contaminated sites. *In:* LINKOV, I. & BAKRRAMADAN, A. (eds.) *Comparative Risk Assessment and Environmental Decision-Making*. Amsterdam: Kluwer Academic Publishers.

LINZ, G. M., HOMAN, H. J., GAULKER, S. M., PENRY, L. B. & BLEIER, W. J. 2007. European starlings: A review of an invasive species with far-reaching impacts. *Managing Vertebrate Invasive Species*, 24.

LIU, S., AURAMBOUT, J.-P., VILLALTA, O., EDWARDS, J., DE BARRO, P., KRITICOS, D. J. & COOK, D. C. 2015. A structured war-gaming framework for managing extreme risks. *Ecological Economics*, 116, 369–377.

LIU, S., PROCTOR, W. & COOK, D. 2010. Using an integrated fuzzy set and deliberative multi-criteria evaluation approach to facilitate decision-making in invasive species management. *Ecological Economics*, 69, 2374–2382.

LODGE, D., WILLIAMS, S., MACISAAC, H., HAYES, K., LEUNG, B., REICHARD, S., MACK, R., MOYLE, P., SMITH, M., ANDOW, D., CARLTON, J. & MCMICHAEL, A. 2006. Biological invasions:

Recommendations for U.S. policy and management. *Ecological Applications*, 16, 2035–2054.

LOW CHOY, S., O'LEARY, R. & MENGERSEN, K. 2009. Elicitation by design in ecology: Using expert opinion to inform priors for Bayesian statistical models. *Ecology*, 90, 265–277.

MACHARIA, J. K. & WANYERA, R. 2012. Effect of stem rust race Ug99 on grain yield and yield components of wheat cultivars in Kenya. *Journal of Agricultural Science and Technology*, 2, 423–431.

MALIPATIL, M. & PLANT HEALTH AUSTRALIA 2008. Threat specific contingency plan: Barley stem gall midge *Mayetiola hordei*. *Industry Biosecurity Plan for the Grains Industry*. Canberra: Plant Health Australia.

MARTIN, T. G., BURGMAN, M. A., FIDLER, F., KUHNERT, P. M., LOW-CHOY, S., MCBRIDE, M. & MENGERSEN, K. 2012. Eliciting expert knowledge in conservation science. *Conservation Biology*, 26, 29–38.

MASSAM, B. 1988. Multi-criteria decision making techniques in planning. *In:* DIAMOND, D. & MCLOUGHLIN, J. (eds.) *Progress in Planning*. Oxford: Pergamon Press.

McCANN, K., HASTINGS, A., HARRISON, S. & WILSON, W. 2000. Population outbreaks in a discrete world. *Theoretical Population Biology*, 57, 97–108.

MCCOSKER, T., MCLEAN, D. & HOLMES, P. 2010. *Northern Beef Situation Analysis 2009*, Sydney: Meat and Livestock Australia.

MCELWEE, H. 2000. The Potential Economic Impact of Khapra Beetle (*Trogoderma granarium* Everts) on the Western Australian Wheat Industry. Bunbury: Government of Western Australia — Department of Agriculture.

MCGAVIN, M. D. 1969. Rubber-vine (*Cryptostegia grandiflora*) toxicity for ruminants. *Queensland Journal of Agricultural and Animal Sciences*, 26, 9.

MCKELVIE, L. 1991. The Economic Impact of Whirling Disease on the Australian Salmonid Industry. Canberra: Australian Bureau of Agricultural and Resource Economics.

MCKELVIE, L., REID, C. & HAQUE, M. 1994. Economic Impact of Salmonid Diseases: Furunculosis and Infectious Haematopoietic Necrosis (IHN), Report to the Australian Quarantine and Inspection Service. Canberra: Australian Bureau of Agricultural and Resource Economics.

MELBOURNE, B. A. & HASTINGS, A. 2009. Highly variable spread rates in replicated biological invasions: Fundamental limits to predictability. *Science*, 325, 1536–1539.

MENDOZA, G. A. & MARTINS, H. 2006. Multi-criteria decision analysis in natural resource management: A critical review of methods and new modelling paradigms. *Forest Ecology and Management*, 230, 1–22.

MERZ, B. & THIEKEN, A. H. 2005. Separating natural and epistemic uncertainty in flood frequency analysis. *Journal of Hydrology*, 309, 114–132.

MO, J., TREVIÑO, M. & PALMER, W. A. 2000. Establishment and distribution of the rubber vine moth, *Euclasta whalleyi* Popescu-Gorj and Constantinescu (Lepidoptera: Pyralidae), following its release in Australia. *Australian Journal of Entomology*, 39, 344–350.

MOODY, M. E. & MACK, R. N. 1988. Controlling the spread of plant invasions: The importance of nascent foci. *Journal of Applied Ecology*, 25, 1009–1021.

MOORE, J. H. & DODD, J. 2008. Eradication of three-horned bedstraw (Galium tricornutum) in Western Australia. *In:* VAN KLINKEN, R. D., OSTEN, V. A., PANETTA, F. D. & SCANLAN, J. C. (eds.) *16th Australian Weeds Conference.* Cairns, Australia: Queensland Weeds Society.

MORAN, P. 1984. *An Introduction to Probability Theory,* Oxford. Clarendon Press.

MORIN, R. S., LIEBHOLD, A. M., TOBIN, P. C., GOTTSCHALK, K. W. & LUZADER, E. 2007. Spread of beech bark disease in the eastern United States and its relationship to regional forest composition. *Canadian Journal of Forest Research,* 37, 726–736.

MORRISON, M. D., BENNETT, J. W., BLAMEY, R. K. & LOUVIERE, J. J. 2002. Choice modelling and tests of benefit transfer. *American Journal of Agricultural Economics*, 84, 161–170.

MUMFORD, J. D. 2001. Environmental risk evaluation in quarantine decision making. *In:* K. ANDERSON, C. M., D. WILSON (ed.) *The Economics of Quarantine and the SPS Agreement.* Adelaide: Centre for International Economic Studies and the Department of Agriculture, Fisheries and Forestry — Australia/Biosecurity Australia.

MUMFORD, J. D., KNIGHT, J. D., COOK, D. C., QUINLAN, M. M., PLUSKE, J. & LEACH, A. W. 2001. *Benefit Cost Analysis of Mediterranean Fruit Fly Management Options in Western Australia,* Ascot: Imperial College.

MUNDA, G., NIJKAMP, P. & RIETVELD, P. 1994. Qualitative multicriteria evaluation for environmental management. *Ecological Economics*, 10, 97–112.

MUNDINGER, P. C. & HOPE, S. 1982. Expansion of the winter range of the house finch: 1947–1979. *American Birds*, 36, 347–353.

MURRAY, G. M. & BRENAN, J. P. 1998. The risk to Australia from *Tilletia indica*, the cause of Karnal bunt of wheat. *Australasian Plant Pathology*, 27, 212–225.

NASH, D. R., AGASSIZ, D. J. L., GODFRAY, H. C. J. & LAWTON, J. H. 1995. The pattern of spread of invading species: Two leaf-mining moths colonizing Great Britain. *Journal of Animal Ecology*, 64, 225–233.

NAYLOR, R. L. 2000. The economics of alien species invasions. *In:* MOONEY, H. A. & HOBBS, R. J. (eds.) *Invasive Species in a Changing World.* Washington, D.C.: Island Press.

NAZARI, K., MAFI, M., YAHYAOUI, A., SINGH, R. P. & PARK, R. F. 2009. Detection of wheat stem rust (*Puccinia graminis* f. sp. *tritici*) race TTKSK (Ug99) in Iran. *Plant Disease*, 93, 317–317.

NEW SOUTH WALES ENVIRONMENT PROTECTION AUTHORITY 2003. Environmental Compliance Report — Wood Preservation Industry: Part A, Compliance Audit. Sydney: New South Wales Environment Protection Authority.

NORTON, B. G. 1998. Improving ecological communication: The role of ecologists in environmental policy formation. *Ecological Applications*, 8, 350–364.

NSW DEPARTMENT OF PRIMARY INDUSTRIES 2015. *NSW Weeds Action Program 2015–2020 Guidelines*. Sydney: NSW Department of Primary Industries.

OERKE, E. C. 2006. Crop losses to pests. *Journal of Agricultural Science*, 144, 31–43.

OFFICE OF PESTICIDE PROGRAMS 2002. Promoting Safety for America's Future. Washington, D.C.: United States Environmental Protection Agency.

OKUBO, A. & LEVIN, S. A. 2002. *Diffusion and Ecological Problems: Modern Perspectives*, New York: Springer.

OSKARSSON, M. C., KLÜTSCH, C. F., BOONYAPRAKOB, U., WILTON, A., TANABE, Y. & SAVOLAINEN, P. 2011. Mitochondrial DNA data indicate an introduction through Mainland Southeast Asia for Australian dingoes and Polynesian domestic dogs. *Proceedings of the Royal Society of London B: Biological Sciences*.

OFFICE OF TECHNOLOGY ASSESSMENT (OTA) 1993. *Harmful Non-Indigenous Species in the United States, OTA-F-565*. Washington, D.C.: Office of Technology Assessment.

ÖZBERK, İ., ATLı, A., YÜCEL, A., ÖZBERK, F. & COŞKUN, Y. 2005. Wheat stem sawfly (*Cephus pygmaeus* L.) damage; impacts on grain yield, quality and marketing prices in Anatolia. *Crop Protection*, 24, 1054–1060.

PALM, M. E. & ROSSMAN, A. Y. 2003. Invasion pathways of terrestrial plant-inhabiting fungi. *In:* RUIZ, G. M. & CARLTON, J. T. (eds.) *Bioinvasions: Pathways, Vectors, and Management Strategies*. New York: Island Press.

PARLIAMENT OF WESTERN AUSTRALIA 2007. Biosecurity and Agriculture Management Act 2007. Western Australia: Department of Premier and Cabinet, State Law Publisher.

PARSONS, W. T. & CUTHBERTSON, E. G. 1992. *Noxious Weeds of Australia*, Melbourne: Inkata Press.

PASEK, J. E. 1998. Khapra Beetle (*Trogoderma granarium* Everts): Pest-Initiated Pest Risk Assessment. Raleigh, NC: United States Department of Agriculture — Animal and Plant Health Inspection Service.

PATERSON, J. & WILKINSON, I. 2015. *Western Australian Canola Industry* [Online]. South Perth: Department of Agriculture and Food, Western Australia. Available at: https://www.agric.wa.gov.au/canola/western-australian-canola-industry [Accessed 4th February 2015].

PATON, D. C., SINCLAIR, R. G. & BENTZ, C. M. 2005. Ecology and management of the common starling (*Sturnus vulgaris*) in the McLaren Vale region. *Final Report to Grape and Wine Research and Development Corporation*. Adelaide: University of Adelaide.

PAYNTER, Q. 2005. Evaluating the impact of a biological control agent Carmenta mimosa on the woody wetland weed *Mimosa pigra* in Australia. *Journal of Applied Ecology*, 42, 1054–1062.

PENNING-ROWSELL, E., JOHNSON, C. & TUNSTALL, S. 2006. 'Signals' from pre-crisis discourse: Lessons from UK flooding for global environmental policy change? *Global Environmental Change-Human and Policy Dimensions*, 16, 323–339.

PERIYANNAN, S., MOORE, J., AYLIFFE, M., BANSAL, U., WANG, X., HUANG, L., DEAL, K., LUO, M., KONG, X. & BARIANA, H. 2013. The gene Sr33, an ortholog of barley Mla genes, encodes resistance to wheat stem rust race Ug99. *Science*, 341, 786–788.

PETERSON, G. D., CARPENTER, S. R. & BROCK, W. A. 2003. Uncertainty and the management of multistate ecosystems: An apparently rational route to collapse. *Ecology*, 84, 1403–1411.

PHELOUNG, P. 2003. An Australian perspective on the management of pathways for invasive species. *In:* RUIZ, G. M. & CARLTON, J. T. (eds.) *Invasive Species Vectors and Management Strategies*. Washington, DC: Island Press.

PHELOUNG, P. C., WILLIAMS, P. A. & HALLOY, S. R. 1999. A weed risk assessment model for use as a biosecurity tool evaluating plant introductions. *Journal of Environmental Management*, 57, 239–251.

PIELKE, R. A., JR. 2007. *The Honest Broker: Making Sense of Science in Policy and Politics*, Cambridge: Cambridge University Press.

PIMENTEL, D. 2014. *Biological Invasions: Economic and Environmental Costs of Alien Plant, Animal, and Microbe Species*, Boca Raton: CRC Press.

PIMENTEL, D., LACH, L., ZUNIGA, R. & MORRISON, D. 2000. Environmental and economic costs associated with non-indigenous species in the US. *BioScience*, 50, 53–65.

PIMENTEL, D., MCNAIR, S., JANECKA, J., WIGHTMAN, J., SIMMONDS, C., O'CONNELL, C., WONG, E., RUSSEL, L., ZERN, J., AQUINO, T. & TSOMONDO, T. 2002. Economic and environmental threats of alien plant, animal and microbe invasions. *In:* PIMENTEL, D. (ed.) *Biological Invasions: Economic and Environmental Costs of Alien Plant, Animal and Microbe Species*. London: CRC Press.

PLANT HEALTH AUSTRALIA 2005. *Government and Plant Industry Cost Sharing Deed in Respect of Emergency Plant Pest Responses*. Canberra: Plant Health Australia.

PORTER, S. D., DE SÁ, L. A. N. & MORRISON, L. W. 2004. Establishment and dispersal of the fire ant decapitating fly *Pseudacteon tricuspis* in North Florida. *Biological Control* 29, 179–188.

PROCTOR, W. 2005. MCDA and Stakeholder participation: Valuing forest resources. *In:* GETZNER, M., SPASH, C. & STAGL, S. (eds.) *Alternatives for Environmental Valuation*. Abingdon: Routledge.

PROCTOR, W. & DRECHSLER, M. 2006. Deliberative multicriteria evaluation. *Environment and Planning C: Government and Policy*, 24, 169–190.

PYŠEK, P. & RICHARDSON, D. M. 2010. Invasive species, environmental change and management, and health. *Annual Review of Environment and Resources*, 35, 25–55.

PYSEK, P., RICHARDSON, D. M., PERGIL, J., JAROSIK, V., SIXTOVA, Z. & WEBER, E. 2008. Geographical and taxonomic biases in invasion ecology. *Trends in Ecology & Evolution*, 23, 237–244.

RAIDAL, S., MCELNEA, C. & CROSS, G. 1993. Seroprevalence of psittacine beak and feather disease in wild psittacine birds in New South Wales. *Australian Veterinary Journal*, 70, 137–139.

RAINBOW LORIKEET WORKING GROUP 2008. Western Australian Rainbow Lorikeet Management Strategy. South Perth: Department of Agriculture and Food, Western Australia.

RANDALL, R. P. 2002. *A Global Compendium of Weeds*, Melbourne: R.G. and F.J. Richardson.

RANGELANDS NRM. 2013. *99 per cent of Rubber vine eradicated on the lower Fitzroy River* [Online]. Como: Rangelands NRM Western Australia. Available at: http://www.rangelandswa.com.au/newsletter.aspx?id=60 [Accessed 3 July 2014].

RAUSCHMAYER, F. & WITTMER, H. 2006. Evaluating deliberative and analytical methods for the resolution of environmental conflicts. *Land Use Policy*, 23, 108–122.

REEVES, A., BAILEY, K. & VINNICOMBE, T. L. 2014. Draft eradication plan for gamba grass (*Andropogon gayanus*) in Western Australia. Bunbury: Department of Agriculture and Food, Western Australia.

REGAN, H. M., BEN-HAIM, Y., LANGFORD, B., WILSON, W. G., LUNDBERG, P., ANDELMAN, S. J. & BURGMAN, M. A. 2005. Robust decision-making under severe uncertainty for conservation management. *Ecological Applications*, 15, 1471–1477.

RENN, O. 2003. The challenge of integrating deliberation and expertise. *In:* MCDANIELS, T. & SMALL, M. J. (eds.) *Risk Analysis and Society: An Interdisciplinary Characterization of the Field*. Cambridge, UK.: Cambridge University Press.

RICCIARDI, A. & COHEN, J. 2007. The invasiveness of an introduced species does not predict its impact. *Biological Invasions*, 9, 309–315.

RICHARDSON, D. M., ALLSOPP, N., D'ANTONIO, C. M., MILTON, S. J. & REJMANEK, M. 2000. Plant invasions — the role of mutualisms. *Biological Reviews*, 75, 65–93.

RIES, D. T. 1926. A biological study of *Cephus pygmaeus* (Linnaeus), the wheat-stem sawfly. *Journal of Agricultural Research*, 33, 277–295.

ROBERTSON, D. P. & HULL, R. B. 2003. Public ecology: An environmental science and policy for global society. *Environmental Science & Policy*, 6, 399–410.

ROSSITER-RACHOR, N. A., SETTERFIELD, S. A., DOUGLAS, M. M., HUTLEY, L. B., COOK, G. D. & SCHMIDT, S. 2009. Invasive Andropogon gayanus (gamba grass) is an ecosystem transformer of nitrogen relations in Australian savanna. *Ecological Applications*, 19, 1546–1560.

ROSSITER, N., SETTERFIELD, S., DOUGLAS, M., HUTLEY, L. & COOK, G. 2002. Exotic grass invasion in the tropical savanna of northern Australia: Ecosystem consequences. *Fuel*, 168–171.

ROY, B. 1985. *Méthodologie multicritere d' aide à la decision,* Paris, Roy, B. (1985). Multi-criteria decision support methodology.

RUIZ, G. M., FOFONOFF, P., HINES, A. H. & GROSHOLZ, E. D. 1999. Non-indigenous species as stressors in estuarine and marine communities: Assessing invasion impacts and interactions. *Limnology and Oceanography,* 44.

RURAL SOLUTIONS SA 2014. Farm gross margin and enterprise planning guide: A gross margin template for crop and livestock enterprises. Adelaide: Grains Research and Development Corporation.

SAAYMAN, M., KRUGELL, W. F. & SAAYMAN, A. 2016. Willingness to pay: Who are the cheap talkers? *Annals of Tourism Research,* 56, 96–111.

SAREWITZ, D. 2004. How science makes environmental controversies worse. *Environmental Science & Policy,* 7, 385–403.

SCANLAN, J., BERMAN, D. & GRANT, W. 2006. Population dynamics of the European rabbit (*Oryctolagus cuniculus*) in north eastern Australia: Simulated responses to control. *Ecological Modelling,* 196, 221–236.

SCANLAN, J. C. & VANDERWOUDE, C. 2006. Modelling the potential spread of *Solenopsis invicta* Buren (Hymenoptera: Formicidae)(red imported fire ant) in Australia. *Australian Journal of Entomology,* 45, 1–9.

SCIENTIFIC COMMITTEE ON TOXICITY, E. A. T. E. 1998. Opinion on the Report by WS Atkins International Ltd. (Vol. B) "Assessment of the Risks to Health and to the Environment of Arsenic in Wood Preservatives and of the Effects of Further Restrictions on its Marketing and Use". *5th Scientific Committee on Toxicity, Ecotoxicity and the Environment Plenary Meeting.* Brussels: European Commission.

SETTERFIELD, S. A., ROSSITER-RACHOR, N. & AMOUZANDEH, I. 2009. The impact of Gamba grass (*Andropogon gayanus*) invasion on the Fire Danger Index (FDI) in Coomalie Shire, Northern Territory. Darwin: Charles Darwin University.

SEWPAC 2011. Feral European Rabbit (*Oryctolagus cuniculus*). Canberra: Department of Sustainability, Environment, Water, Population and Communities.

SHARMA, I., BAINS, N. S. & SHARMA, R. C. 2012. Resistance in wheat to Karnal bunt. *In:* SHARMA, I. (ed.) *Disease Resistance in Wheat.* Wallingford, UK: CABI International.

SHIGESADA, N. & KAWASAKI, K. 1997. *Biological Invasions: Theory and Practice,* Oxford: Oxford University Press.

SHIGESADA, N., KAWASAKI, K. & TAKEDA, Y. 1995. Modeling stratified diffusion in biological invasions. *The American Naturalist,* 146, 229–251.

SIMBERLOFF, D. 2006. Risk assessments, blacklists, and white lists for introduced species: Are predictions good enough to be useful? *Agricultural and Resource Economics Review,* 35, 1–10.

SIMBERLOFF, D. & VON HOLLE, B. 1999. Positive interactions of nonindigenous species: Invasional meltdown? *Biological Invasions,* 1, 21–32.

SINGH, R. P., HODSON, D. P., HUERTA-ESPINO, J., JIN, Y., BHAVANI, S., NJAU, P., HERRERA-FOESSEL, S., SINGH, P. K., SINGH, S. & GOVINDAN, V. 2011. The emergence of Ug99 races of the stem rust fungus is a threat to world wheat production. *Annual Review of Phytopathology*, 49, 465–481.

SKELLAM, J. G. 1951. Random dispersal in theoretical populations. *Biometrika*, 38, 196–218.

SMITH, C. S., LONSDALE, W. M. & FORTUNE, J. 1999. When to ignore advice: Invasion predictions and decision theory. *Biological Invasions*, 1, 89–96.

STANSBURY, C. D. & PRETORIUS, Z. A. 2001. Modelling the potential distribution of Karnal bunt of wheat in South Africa. *South African Journal of Plant Soil*, 18, 159–168.

STOKSTAD, E. 2007. Deadly wheat fungus threatens world's breadbaskets. *Science*, 315, 1786–1787.

STRAYER, D. L., EVINER, V. T., JESCHKE, J. M. & PACE, M. L. 2006. Understanding the long-term effects of species invasions. *Trends in Ecology & Evolution*, 21, 645–651.

STRIVE, T., WRIGHT, J., KOVALISKI, J., BOTTI, G. & CAPUCCI, L. 2010. The non-pathogenic Australian lagovirus RCV-A1 causes a prolonged infection and elicits partial cross-protection to rabbit haemorrhagic disease virus. *Virology*, 398, 125–134.

THE ALLEN CONSULTING GROUP 2006. Draft: Building Regulations for European House Borer Regulatory Impact Statement, Report to the Department of Housing and Works, Perth.

THOMAS, M. B. & REID, A. M. 2007. Are exotic natural enemies an effective way of controlling invasive plants? *Trends in Ecology and Evolution*, 22, 447–453.

THOMPSON, H. V. & KING, C. M. 1994. *The European Rabbit: The History and Biology of a Successful Colonizer*, Oxford: Oxford University Press.

THORNE, F. S., BRENNAN, J. P., KELLY, P. W. & KINSELLA, A. 2004. Evaluating the Economic Impact of an Outbreak of a Quarantinable Disease in Europe: The Case of an Outbreak of Karnal Bunt of Wheat in the EU. *78th Annual Conference of the Agricultural Economics Society*. South Kensington: Imperial College London.

TOBIN, P. C., LIEBHOLD, A. M. & ANDERSON ROBERTS, E. 2007. Comparison of methods for estimating the spread of a non-indigenous species. *Journal of Biogeography*, 34, 305–312.

TOMLEY, A. & EVANS, H. 2004. Establishment of, and preliminary impact studies on, the rust, *Maravalia cryptostegiae*, of the invasive alien weed, *Cryptostegia grandiflora* in Queensland, Australia. *Plant Pathology*, 53, 475–484.

TOMLEY, A. J. 1995. The biology of Australian weeds. 26. *Cryptostegia grandiflora* R. Br. *Plant Protection Quarterly*, 10, 122–130.

TRACEY, J., BOMFORD, M., HART, Q., SAUNDERS, G. & SINCLAIR, R. 2007. Managing Bird Damage to Fruit and Other Horticultural Crops. Canberra: Bureau of Rural Sciences.

TROPE, Y., LIBERMAN, N. & WAKSLAK, C. 2007. Construal levels and psychological distance: Effects on representation, prediction, evaluation, and behavior construal level theory. *Journal of Consumer Psychology*, 17, 83–95.

ULMER, B. J. & DOSDALL, L. M. 2006. *Spring Emergence Biology of the Cabbage Seedpod Weevil (Coleoptera: Curculionidae)*. Annals of the Entomological Society of America, 99, 64–69.

ULUBASOGLU, M., MALLICK, D., WADUD, M., HONE, P. & HASZLER, H. 2011. How price affects the demand for food in Australia — An analysis of domestic demand elasticities for rural marketing and policy. Canberra: Rural Industries Research and Development Corporation.

VAN KLEUNEN, M., DAWSON, W., ESSL, F., PERGL, J., WINTER, M., WEBER, E., KREFT, H., WEIGELT, P., KARTESZ, J., NISHINO, M., ANTONOVA, L. A., BARCELONA, J. F., CABEZAS, F. J., CARDENAS, D., CARDENAS-TORO, J., CASTANO, N., CHACON, E., CHATELAIN, C., EBEL, A. L., FIGUEIREDO, E., FUENTES, N., GROOM, Q. J., HENDERSON, L., INDERJIT, KUPRIYANOV, A., MASCIADRI, S., MEERMAN, J., MOROZOVA, O., MOSER, D., NICKRENT, D. L., PATZELT, A., PELSER, P. B., BAPTISTE, M. P., POOPATH, M., SCHULZE, M., SEEBENS, H., SHU, W.-S., THOMAS, J., VELAYOS, M., WIERINGA, J. J. & PYSEK, P. 2015. Global exchange and accumulation of non-native plants. *Nature*, 525, 100–103.

VINSON, S. B. 1997. Invasion of the red imported fire ant (Hymenoptera: Formicidae): spread, biology, and impact. *American Entomologist*, 43, 23–39.

WAAGE, J. K., FRASER, R. W., MUMFORD, J. D., COOK, D. C. & WILBY, A. 2005. A New Agenda for Biosecurity *Horizon Scanning Programme*. London: Department for Food, Environment and Rural Affairs.

WAAGE, J. K. & MUMFORD, J. D. 2008. Agricultural biosecurity. *Philosophical Transactions of the Royal Society of London B: Biological Sciences*, 363, 863–876.

WEBB, T. J. & RAFFAELLI, D. 2008. Conversations in conservation: Revealing and dealing with language differences in environmental conflicts. *Journal of Applied Ecology*, 45, 1198–1204.

WEISS, M. J. & MORRILL, W. L. 1992. Wheat stem sawfly (Hymenoptera: Cephidae) revisited. *American Entomologist*, 38, 241–245.

WELLINGS, C. & PLANT HEALTH AUSTRALIA 2010. Threat Specific Contingency Plan: Barley stripe rust (*Puccinia striiformis* f. sp. *hordei*). *Industry Biosecurity Plan for the Grains Industry*. Canberra: Plant Health Australia.

WELLINGS, C. R., BURDON, J. J., MCINTOSH, R. A., WALLWORK, H., RAMAN, H. & MURRAY, G. M. 2000. A new variant of *Puccinia striiformis* causing stripe rust on barley and wild *Hordeum* species in Australia. *Plant Pathology*, 49, 803–803.

WELLINGS, C. R., MCINTOSH, R. A. & WALKER, J. 1987. *Puccinia striiformis* f. sp. *tritici* in Eastern Australia possible means of entry and implications for plant quarantine. *Plant Pathology*, 36, 239–241.

WHITBY, M. 2000. Challenges and options for the UK agri-environment: Presidential address. *Journal of Agricultural Economics*, 51, 317–332.

WHITE, S., FANE, S. A., GIURCO, D. & TURNER, A. J. 2008. Putting the economics in its place: Decision-making in an uncertain environment. *In:* ZOGRAFOS, C. & HOWARTH, R. B. (eds.) *Deliberative Ecological Economics*. London: Oxford University Press.

WILLIAMSON, M. 1996. *Biological Invasions*, London: Chapman and Hall.

WOOLNOUGH, A., MASSAM, M., PAYNE, R. & PICKLES, G. 2005 Out on the border: Keeping starlings out of Western Australia. *13th Australasian Vertebrate Pest Conference*, Wellington, New Zealand, 2–6 May, 2005. Manaaki Whenua Press, Landcare Research, 183–189.

WTO 2000. Australia — Measures Affecting Importation Of Salmon — Recourse To Article 21.5 By Canada — Report Of The Panel. Geneva: World Trade Organisation.

YEH, C.-H. & CHANG, Y.-H. 2009. Modeling subjective evaluation for fuzzy group multicriteria decision making. *European Journal of Operational Research*, 194, 464–473.

YEH, C.-H., DENG, H. & CHANG, Y.-H. 2000. Fuzzy multicriteria analysis for performance evaluation of bus companies. *European Journal of Operational Research*, 126, 459–473.

ZADEH, L. A. 1965. Fuzzy sets. *Information and Control*, 8, 338–353.

ZADOCKS, J. & SHEIN, R. 1979. *Epidemiology and Plant Disease Management*, Oxford: Oxford University Press.

ZELENY, M. 1982. *Multiple Criteria Decision Making*, New York: McGraw-Hill.

Index